# GOING TO MARS

The Stories of the People Behind NASA's

Mars Missions Past, Present, and Future

Brian Muirhead

Flight System Manager, Mars *Pathfinder*

*and* Judith & Garfield Reeves-Stevens

POCKET BOOKS

New York  London  Toronto  Sydney

POCKET BOOKS, a division of Simon & Schuster, Inc.
1230 Avenue of the Americas, New York, NY 10020

ISBN: 0-671-02796-4
First Pocket Books trade paperback edition November 2004

10  9  8  7  6  5  4  3  2  1

POCKET and colophon are registered trademarks of Simon & Schuster, Inc.

Book design by Richard Oriolo

Manufactured in the United States of America

For information regarding special discounts for bulk purchases, please contact Simon & Schuster Special Sales at 1-800-456-6798 or business@simonandschuster.com

## DEDICATION

The first person to set foot on Mars is alive today,

a girl or boy who might just be finishing grade school,

watching *Star Trek* on TV, conducting messy science experiments in the kitchen,

sketching spaceships in notebooks, and maybe, late at night,

dreaming of what it might be like to live those imaginary adventures for real

To that child, whoever, wherever, he or she might be today,

as a record of the first steps taken by those

who share the same dreams,

we dedicate this book.

# CONTENTS

Risk is our business.
    —CAPTAIN JAMES T. KIRK

# FOREWORD

WHO ARE WE? WHERE DID we come from? Are we alone? What will be our fate?

More than anything else, the search for answers to those questions has inspired philosophy and science since the first cultures arose. Yet it is only this generation, in the dawn of the twenty-first century, that has acquired the technology that can finally help reveal some definitive answers. But not here on Earth—on Mars.

I have been privileged to work at the Jet Propulsion Laboratory for more than three decades. In 2001, I became its director. Through all those years and all the projects I have taken part in, as a scientist, engineer, co-investigator, principal investigator, and now as the Lab's director, every mission that JPL has launched has in some way been intended to further our understanding of our world, our solar system, and the universe, and thus further our understanding of ourselves.

Of those many missions, certainly the extraordinarily successful Mars Exploration Rovers, *Spirit* and *Opportunity*, are among the most well known. The challenge of developing and launching two rovers to Mars in three years was nothing short of daunting. But that is what exploration is all about: commit to mighty goals, assemble a dream team, and undertake the journey of exploration even though the shadow of failure is always around the corner. Setbacks were encountered numerous times during the development of *Spirit* and *Opportunity*. Some of these setbacks seemed sometimes to be insurmountable. But the team at JPL kept focused, stayed confident, and applied their innovative spirit to overcome all problems, leading to the success of these missions.

As director of JPL, I am gratified and inspired by the public's enthusiasm for the Mars Exploration Rover mission (or MER). As a researcher, I am impressed by my colleagues at the Lab, at NASA, and at all the institutions and private contractors who contributed to the mission's groundbreaking accomplishments—which *still* continue as I write these words, far longer than either rover was required to operate.

In these pages, the reader will be able to meet some of my colleagues—the engineers and scientists who took part in this historic undertaking, in addition to those who laid the groundwork for *Spirit* and *Opportunity* with the Mars Pathfinder mission of 1997, which carried the first Mars rover, *Sojourner*. These successes have been preceded and supported by other groundbreaking missions, starting with *Mariner 4*, and including *Viking*, Mars Global Surveyor, and Mars Odyssey.

No doubt other books have and will be written to describe the technology and explain the science of MER and Pathfinder, but in this book the authors have told the human side of the story. Engineers and scientists aren't stereotypical computers on legs—they're grade-school kids inspired by a fanciful story the teacher once read about Dr. Dolittle's trip to the Moon; an eight-year-old astronomer looking through an Edmund Scientific telescope for the very first time to see the moons of Jupiter just as Galileo did; a farm boy who read every book he could get about building his own radio, though his family couldn't afford to buy the parts; a young girl planning to be a concert pianist, but inspired to become an engineer by the photograph of a rocket hanging above her parents' stereo.

Above all else, what this book makes clear is that going to Mars is a human endeavor. I believe it is this human connection that ties us all so strongly to the excitement of exploring Mars, whether scientist, engineer, or interested member of the public. That excitement exists, in part, because of what we all hope we might find there—the answers to our four questions.

The planets, and especially Mars, are laboratories where we can conduct investigations not possible on Earth. The steps by which our planet changed from being a world of chemical reactions to one of biology are lost to us. Earth is a dynamic planet, and the rocks that existed when life first formed here have been extensively modified by geological processes.

But Mars is a less active world, and if life formed there in a manner and at a time similar to life on Earth, even at a cellular level, then the record of its emergence might still be preserved in Martian rocks for us to read.

Even more compelling, and exciting, is the possibility that if life did arise on Mars, and was similar to life on Earth, then it might be that some form of that life could still exist, perhaps somewhere beneath the planet's surface.

Imagine what the discovery of life on Mars, in the past or present, would mean to our understanding of life in the universe—a universe with billions of galaxies each containing hundreds of millions of stars, many of which have planets. If *two* planets in our solar system gave rise to life—either independently or because of interplanetary seeding from one to another or even from a third source—what might that tell us of the likelihood of other worlds with other life-forms throughout the universe?

But then imagine the profound effect that would come from exploring Mars and determining that life had *never* arisen there, even when it was warm and wet and subject to "contamination" from Earth bacteria carried there by meteorites.

As noted science-fiction writer Sir Arthur Clarke once said, "Sometimes I think we're alone in the universe, and sometimes I think we're not. In either case, the idea is quite staggering."

Who are we? Where did we come from? Are we alone? What will be our fate?

Our search for the answers to these questions is nothing less than the search for understanding the origin of life in our solar system and beyond.

If the answers truly are waiting for us on Mars, then the people whose stories are told in this book are among those who have helped us take the first steps in the most important mission of exploration our generation can undertake. I hope you enjoy the journey as much as I do.

—Dr. Charles Elachi
Director, Jet Propulsion Laboratory
August 2004

# Between WHAT IS and WHAT IF

## The Story of Mars Told Two Ways

*Science and science fiction have done a kind of dance over the last century, particularly with respect to Mars. The scientists make a finding, it inspires science-fiction writers to write about it, and a host of young people read the science fiction and are excited and inspired to become scientists to find out more about Mars, which they do, which then feeds again into another generation of science fiction and science.*

*That sequence has played a major role in our present capability to get to Mars.*

—CARL SAGAN

WE ARE ALL GOING TO Mars, and this book tells the story of our first steps on that journey in a way not done before, because our story is about more than exploration. It is also about the people who are leading us in that great adventure. These explorers, like their individual missions, differ in style, rhythm, energy, and performance. The single goal that unites them all in Carl Sagan's interactive dance of many steps by many skilled people from many different backgrounds, is their overwhelming drive to achieve perfection. Because nothing less will get us to Mars. And beyond.

One half of our writing team—Brian Muirhead—comes from the real world of science and engineering: NASA and the Jet Propulsion Laboratory. With his own hands he has built hardware that has flown to Jupiter. He helped assemble and lead the team that electrified the world by landing the first rover on Mars in 1997. Then, as chief engineer of Mars Science Laboratory (also known as the MSL), he helped develop a new generation of rovers proposed to land on Mars in 2009. These new rovers will begin to blaze the trail for the eventual and inevitable first human explorers. The MSL rover, with the possibility of *years* of operation, will be capable

of roaming miles over rugged terrain almost anywhere on Mars, looking for signs of whether Mars could have harbored life in the past or may even do so now.

The other half of our writing team—Judith and Garfield Reeves-Stevens—comes from the world of science fiction. They've written of aliens and starships, and in some of their novels have imagined Mars as it might be, in a future when colonists study Martian fossils firsthand or fight for their independence from the governments of Earth.

Real science deals with answering the question "What is?"

Science fiction deals with the question "What if?"

In this book, both halves of our writing team combine both their unique perspectives to answer the more specific questions of *how* and *why* and *when* we have gone and surely will continue to go to Mars.

(OPPOSITE PAGE) Brian Muirhead first went to work at the Jet Propulsion Laboratory in 1977, as a summer student. After graduating from the University of New Mexico, he returned to JPL in 1978 as an engineer on the Galileo mission. The *Galileo* orbiter, seen here, launched from the Space Shuttle *Atlantis* in October 1989 and reached Jupiter six years later. After releasing a probe into Jupiter's atmosphere, *Galileo* continued to study Jupiter and its moons until December 2003, when it was deliberately sent into the atmosphere of Jupiter and destroyed in order to avoid the possibility that it might one day crash and contaminate Jovian moons that might harbor life. COURTESY NASA.

In 1996, while Brian Muirhead was working to send the Pathfinder mission to Mars, Judith and Garfield Reeves-Stevens were working in the more fanciful realms of science fiction. COURTESY SMARTSHOW ENTERTAINMENT.

## DOES ANYBODY REALLY KNOW WHAT TIME IT IS?

*Different Years for Different Reasons*

By definition, the length of an Earth year is the time it takes for the Earth to complete one orbit around the Sun.

For everyday life, most people follow what is called the calendrical year, which has the familiar length of 365 days, with an extra day added almost every four years. (And to be a bit more precise, once each century, for fine adjustment, no leap day is added.)

The extra day is necessary because the Earth actually takes about 365 and a quarter days to complete an orbit. Without adding an extra day almost every time those four quarters add up to 1, our measurement of the seasons would slowly drift so that after only 120 years the calendar would be about a month ahead of the actual position of the Sun, which determines the equinoxes and solstices that define each season.

While 365 full days define the calendrical year, the actual length of time it takes the Earth to complete its orbit— ~365.2425 days—is considered the solar year. However, for some astronomical calculations, scientists must also take into account the fact that as the Earth orbits the Sun, the Sun is moving on its own orbital path around the center of the galaxy. Thus, instead of measuring the Earth's position in relation to the moving Sun, sometimes it's necessary to measure the Earth's position in relation to a "fixed" star—that is, a star so distant that it does not appear to move. The time it takes for the Earth to complete an orbit of the Sun that returns it to the same position in relation to a fixed star is called a sidereal year. Averaged over the years 1994 to 2000, the sidereal year was approximately 365.256363 days, which is slightly longer than the average solar year of 365.2425 over that same period.

The tropical year, which is the year most often used by astronomers, is the time it takes for the Earth to return to a specific equinox—which can be defined as one of two points in its orbit around the Sun in which the axis of the planet's spin is angled at exactly ninety degrees to the sun. The vernal equinox, which occurs around March 21, marks the beginning of spring in the Northern Hemisphere. The autumnal equinox, which occurs around September 23, marks the beginning of fall in the Northern Hemisphere. On both days, the length of day and night are, to all intents and purposes, equal. But to be precise, as scientists are inclined to be, because of the refraction of sunlight through the edges of the Earth's atmosphere, the day is actually slightly longer than the night.

The anomalistic year is the time it takes for the Earth to return to its farthest point from the Sun—called aphelion. Between 1994 to 2000, the anomalistic year was approximately 365.259635, which was slightly longer than the sidereal year.

Aren't you glad you asked?

There's another advantage to having storytellers who, together, can look in from the outside, and out from the inside, at the same time.

Scientists and engineers like Brian, who do the hard, nuts-and-bolts, real work of going to Mars, operate in a world of numbers and precision absolutely critical to the successful performance of their jobs. Yet that world is often confusing to the average person who is neither a scientist nor an engineer.

Here's an example.

Ask the average person the length of the year, and you'll probably get the answer, 365 days. A few people might go on to say that every four years, a leap year has 366 days because of that extra day, the twenty-ninth, that's added to the end of February. Both answers are perfectly adequate for most people and most tasks, and there's nothing wrong with them.

But, for a specialist, they're not *precisely* right.

For instance, if you were to ask an astronomer the length of a year, he or she is very likely to answer with a new question: Which *kind* of year do you mean? The sidereal year? The solar year? The tropical year? How about the anomalistic year?

For the astronomer, if not for the average person, these distinctions are crucial, because there are many different definitions of "a year," and not all of them are the same.

Here's why—and feel free to skip the decimals.

The sidereal year, for example, is about 365.256363 days long. That's 0.013863 days longer than the solar year of 365.2425 days. The difference between the sidereal and solar years amounts to about twenty minutes, which has absolutely no effect on when the rent is due or what time a movie starts on a Friday night. But to a spacecraft navigator calculating the trajectory of a Mars probe, and to the engineer determining the precise angle a rocket engine must tilt and how many seconds it must fire to achieve that trajectory, those twenty minutes can mean the difference between success and looking for a new line of work.

As for those other years, an anomalistic year is about 365.259635 days long. A tropical year—which is the type of year astronomers usually use in their calculations—averaged out to about 365.242190 days over the years 1994 to 2000. Why is it necessary to specify 1994 to 2000? Because the length of all those different types of years—sidereal, solar, anomalistic, and tropical—are not the *exact* same length from one year to the next!

So, there are *two* types of answers to the apparently simple question: "How long is a year?" There's the everyday answer of 365 days, and 366 in a leap year. And then there's the precise scientific answer of . . . well, that one could fill an entire book with historical records, mathematical formulae, and complex calculations, and even then that book would conclude with the important cautionary observation that every measurement of a year's length remains subject to change.

Is it any wonder that scientists and engineers are often viewed by the average person as being incapable of giving a simple answer to a simple question? That's the price they pay for requiring precision and accuracy in their work. Think about it. How often are you presented with a piece of information that is extremely precise but could be wrong (inaccurate) or with a fact that is accurate but so vague (imprecise) as to be useless? Much of what we get through our media suffers from this dilemma. (Engineers and scientists are trained to ask questions, discern and understand the difference, and when it comes to space reporting, they wish reporters were too!) This is why scientists and engineers sometimes cringe when some of their colleagues try to distill the essence of a scientific finding or an engineering principle into something that makes sense to the average person. Even Carl Sagan faced this, and for all the adulation he deservedly received from the public, he also had to deal with criticism from his peers for oversimplifying and not being precise enough.

Two of us—Judith and Garfield Reeves-Stevens—are definitely nonspecialists, what Ed Weiler, NASA's associate administrator for space science, calls "real people." They have no degrees in mathematics and science, have never published papers in distinguished research journals—ask them the length of the year, and they say 365 days, 366 in leap years.

# 1877: A SPACE ODYSSEY

*The Year Mars Changed Forever*

Mars and Earth each orbit the Sun at different speeds and distances, and every two years and seven weeks or so, when the Earth catches up to the slower moving Mars, the two planets make a close approach, which astronomers call an "opposition." (That name is used because, from the Earth's perspective, when an opposition occurs, Mars is directly opposite the Sun.)

Invariably, when an opposition nears and Mars becomes brighter and more noticeable in the night sky, our interest in it grows. Certainly, the extremely close opposition of August 2003 generated unprecedented attention for the Red Planet, especially because at the time five spacecraft were speeding toward it: NASA's triumphant Mars Exploration Rovers, *Spirit* and *Opportunity;* the European Space Agency's successful first Mars orbiter, *Mars Express;* the ill-fated British landing probe, *Beagle 2;* and the ultimately unsuccessful Japanese orbiter, *Nozomi.*

This connection between close approaches and renewed public interest in Mars has been going on for generations, ever since what could be considered Earth's most notable close approach to the Red Planet in August 1877.

Earlier oppositions had reliably fueled astronomers' interest in Mars—the closer Mars and Earth approach each other, the more detail can be seen. Throughout the 1800s, as a result of observations with constantly improving telescopes, scientists' knowledge of Mars slowly increased. The close approach of 1832 enabled the German astronomer Johann Heinrich von Mädler to calculate the length of the Martian day to within a second. During a series of close oppositions from 1858 to 1864, the first color sketches of Martian surface features were made by the Jesuit astronomer Father Pietro Angelo Secchi. And in 1867, British astronomer Richard Anthony Proctor published one of the first detailed maps of Mars. (Unfortunately, Proctor based some of his work on what's considered to be the first map of Mars, drawn by astronomer von Mädler and a German banker, Wilhem Beer, which bears no resemblance to any existing features of Mars other than the polar caps. Though Proctor's map was quickly superseded, his choice of where to draw the zero meridian of Martian longitude is still used today.)

What made the opposition of 1877 so much more momentous than any previous close approach were the observations made by two prominent astronomers: an American, Asaph Hall, and an Italian, Giovanni Virginio Schiaparelli.

Hall was the first to discover the Martian "moons" Phobos and Deimos. But it is Schiaparelli's "discovery" that dramatically altered humanity's vision of Mars, even though modern observations have shown that what Schiaparelli thought he saw was nothing more than equal parts blurry vision and wishful thinking. For when Schiaparelli's observations became known to the world, Mars forever ceased to be just another point of light in the sky.

Instead, almost overnight, Mars became an *inhabited* world, home to intelligent beings who had constructed a planetary system of irrigation canals to bring water from the Martian polar ice caps to its parched equatorial deserts!

During that opposition of August 1877, Schiaparelli had seen on the surface of Mars what he called *"canali."* In context, the Italian word meant "channels," specifically those carved by free-flowing water. In fact, the same word had been used by Father Secchi in 1869 to describe features he believed he saw on Mars. But when Schiaparelli's observations were translated into English, French, and German, *canali* became "canals," as in *artificial* waterways, just like the engineering marvel of the age, the Suez Canal linking the Mediterranean and Red Seas, which had opened to shipping only eight years earlier.

For centuries, scholars had debated a philosophical premise known as the "Plurality of Worlds"—essentially the possibility that intelligent life might exist elsewhere among the planets and stars.

But in August 1877, that question suddenly became more than a philosophical topic. For the first time in human history, not just scholars but ordinary people could look up at one particular light in the sky and have

Ask the other half of the team—Brian Muirhead—the length of the year, and his response not only will be to ask which kind of year you mean, he's liable to ask which planet you mean: "Earth, Mars, Jupiter?"

The goal of this book is to bridge that gap between communication and understanding, by sharing the stories of the people involved in a great adventure in exploration, and by communicating not just the names of things but their underlying meaning—what Richard Feynman, one of the greatest physicists of the twentieth century, called the "real knowledge" from which the laws of science and nature arise.

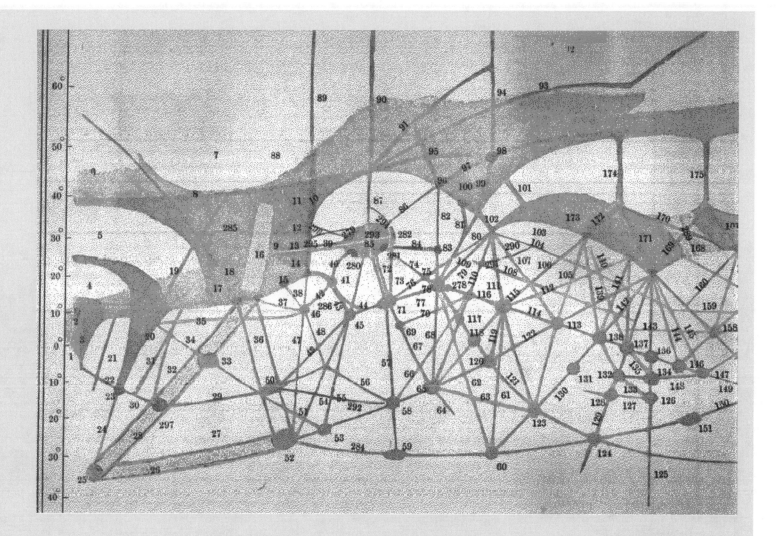

This map of the Martian canals was drawn by American astronomer Percival Lowell and published in his groundbreaking 1895 book simply titled *Mars*. In it, Lowell laid out the logical argument that these extensive waterways could only be artificial, inspiring generations of astronomers, the general public, and science-fiction writers to believe that Mars might actually be inhabited by intelligent beings.

reasonable cause to wonder if anyone was looking back.

Mars—and Earth—would never be the same.

What unites both halves of our odd-couple writing team is the passion we all share: for space exploration, for Mars, and for telling a story that will stir that same passion in all who read it.

Brian Muirhead is proud of the work he has done and is doing. As an active insider-participant in space exploration himself, he is especially proud to be associated with the talented, dedicated, and diverse people he has worked with in the past and continues to work with every day at NASA, the Jet Propulsion Laboratory (JPL), and at numerous private-industry partners and suppliers.

## SETTING NASA'S GOALS

*Excerpts from* The Vision for Space Exploration
*February 2004*

NASA was created on July 29, 1958, when President Dwight D. Eisenhower signed the National Aeronautics and Space Act. Since that time, like any U.S. government agency, NASA has taken its direction from the country's elected leaders. In 1961, President John F. Kennedy charged NASA with the mission to land humans on the Moon and safely return them to Earth before 1970. In 1984, President Ronald Reagan approved the construction of Space Station *Freedom,* which was the first step in the planning and construction of what today is the International Space Station, also known as Space Station *Alpha.*

On January 14, 2004, after almost a year of planning by top NASA personnel, scientists, engineers, and space exploration advocates, President George W. Bush presented NASA's new objective: to focus its efforts on the ultimate goal of sending humans to Mars.

Before that goal can be achieved, there is a great deal of information that NASA needs to obtain about conditions on Mars, and considerable technical expertise that must be developed in order to guarantee the safety of astronauts who could be away from the Earth for as long as three years.

But as the old saying notes, all journeys begin with a single step, and in response to the president's directive, in February 2004, NASA described its newest challenge—and humanity's—in a document titled *The Vision for Space Exploration.*

These are some excerpts.

*"Somewhere, something incredible is waiting to be known."*
—CARL SAGAN

### Goal and Objectives

The fundamental goal of this vision is to advance U.S. scientific, security, and economic interests through a robust space exploration program. In support of this goal, the United States will:

- Implement a sustained and affordable human and robotic program to explore the solar system and beyond;
- Extend human presence across the solar system, starting with a human return to the Moon by the year 2020, in preparation for human exploration of Mars and other destinations; . . .

### Mars Research, Testbeds, and Missions

NASA is aggressively pursuing the search for water and life on Mars using robotic explorers. The *Spirit* and *Opportunity* rovers that landed on Mars in January 2004 are the latest in a series of research missions planned to explore Mars through 2010. By the end of this decade, three rovers, a lander, and two orbiters will have visited the planet. NASA will augment this program and prepare for the next decade of Mars research missions by investing in key capabilities to enable advanced robotic missions, such as returning geological samples from Mars or drilling under the surface of Mars. This suite of technologies will enable NASA to rapidly pursue the search for water and life at Mars wherever it may lead in the next decade.

Starting in 2011, NASA will also launch the first in a new series of human precursor missions to Mars. These robotic testbeds will demonstrate technologies—such as improved aerodynamic entry, Mars orbital rendezvous and docking, precision landing, resource extraction and utilization, and optical communications—that can greatly enhance future robotic capabilities and are key to enabling future human Mars missions. These missions will also obtain critical data for future human missions on chemical hazards, resource locations, and research sites. They may prepare resources and sites in anticipation of human landings . . .

The timing of the first human research missions to Mars will depend on discoveries from robotic explorers, the development of techniques to mitigate Mars hazards, advances in capabilities for sustainable exploration, and available resources.

*"Sometimes I think we're alone in the universe, and sometimes I think we're not. In either case, the idea is quite staggering . . ."*
—ARTHUR C. CLARKE

### National Benefits

Space exploration holds a special place in the human imagination. Youth are especially drawn to Mars rovers, astronauts, and telescopes. If engaged effectively and creatively, space inspires children to seek careers in math, science, and engineering, careers that are critical to our future national economic competitiveness. . . .

When the unknown becomes known, it catalyzes change, stimulating human thought, creativity, and imagination. The scientific questions that this plan pursues have the potential to revolutionize whole fields of research. For example, scientists are still working to understand how similarly sized planets, such as Mars and Earth, could have developed so differently and what that could mean for our planet. If life is found beyond Earth, biological processes on other worlds may be very different from those evolved on our world. Outside the sciences, the very knowledge that life exists elsewhere in the universe may hold revelations for fields in the humanities.

Exploration and discovery are key agents of growth in society—technologically, economically, socially, internationally, and intellectually. This plan sets in motion activities that will contribute to change and growth in the U.S. and the world over the next century.

## ACROSS THE ZODIAC AND ON TWO PLANETS

*Based on an Almost True Story . . .*

The nature of entertainment—its inspiration and methods—hasn't changed much even as entertainment media have rapidly evolved in the past century. It really hasn't changed all that much since the days when storytellers sat around campfires after a hard day hunting and cooking woolly mammoths and told tales of their adventures and disasters, and those of other times and places. So just as no one today is surprised when book publishers and moviemakers rush to produce entertainment based on the latest crazes, it's not remarkable that fiction writers in the past found inspiration in Schiaparelli's Martian canals.

What is considered to be the first novel describing a trip to Mars was written by the British author Percy Gregg and was published three years after the historic opposition of Mars in 1877. *Across the Zodiac* tells the tale of a scientist who harnesses a form of negative gravity called "apergy" to pilot a spacecraft to Mars.

In some ways, the diminutive Martians of the tale are more advanced than humans—most work is done either by machines or by specially bred animals. In other ways,

Gregg's Martians are much like humans—they are so convinced there is no other life in the universe that they refuse to believe the Earth man is from Earth. Instead, they decide he is an oddly tall Martian from a far-off region of their own world.

Other than its descriptions of the main character's early experiments with apergy, *Across the Zodiac* has little in the way of what today's readers would recognize as science fiction. However, the next notable novel about Mars more than made up for that lapse.

*On Two Planets* was published in 1897, after two decades of ongoing observations had "confirmed" the existence of Schiaparelli's canals. This novel, originally published in German as *Auf zwei Planeten,* was written by Kurd Lasswitz, a professor of mathematics who had previously written nonfiction books on the history of science. Just like Isaac Asimov, Gregory Benford, Gentry Lee, and other scientists/engineers turned science-fiction writers of today, Lasswitz's background helped him create a story full of Martian technological advancements befitting a species that had built such advanced waterways.

Lasswitz's story begins with a team of scientists discovering a Martian base at the North Pole, which is serviced by a Martian space station above the Earth. (A second Martian base is located in Antarctica.) While the Martians have fantastically advanced

technology, including an energy projector intended for peaceful purposes but which can still destroy a city with a single blast, they are ethical creatures devoted to peace.

When some of the human scientists accept an invitation to visit Mars, they learn that the vast system of canals are, in fact, real. They also learn that while a small faction of Martians are determined to take over the Earth in order to exploit its more abundant solar energy, another group of Martians want to share their advanced science with humans. Unfortunately, a misunderstanding leads to an outbreak of hostilities between the worlds, and the Martian spaceships easily defeat Earth's armies. However, the Martians get bogged down in trying to occupy and rule the Earth, and when secret factories in the United States succeed in building duplicates of the Martian spaceships, the resulting standoff finally leads to peace between the planets.

*On Two Planets* was the science-fiction blockbuster of its day, devoured by readers around the world who were eager to read about what might be waiting on Mars.

But in the same year that novel was published in Germany, another tale of Mars was being serialized in England's *Pearson's* magazine. And though few today know anything about Lasswitz's epic story, that other 1897 novel about Mars and Martians continues to live in the public's imagination—H. G. Wells's *The War of the Worlds.*

As science-fiction writers and space enthusiasts, Judith and Garfield Reeves-Stevens are deeply appreciative of the amazing work that NASA, and JPL, and the aerospace industry accomplish. And as outsider-observers—members of the everyday, taxpaying public—they're also sometimes critical, and raise points and ask questions that may not occur to insiders.

Both our perspectives—observer and participant—join together in this book to tell a story that both the average person and the specialist can enjoy.

Here's what you'll find in these pages.

First and foremost, all readers will find in the main text the most important component of our journey to Mars—the story of the *people* who are working to take

## THE WAR OF THE WORLDS

*The Martian Novel*

Not only was *The War of the Worlds* the first novel published in English to present Mars and Martians in scientific terms, it was the very first outright alien-invasion story—one that has set the standard for more than a century of science fiction. For the first time aliens were monstrous, plus they used advanced-technology weaponry: heat rays that burned whatever they touched and a poison gas called "Black Smoke" that pre-dated the use of gas in World War I by al-most twenty years. Even more impressively, the opening lines of this seminal novel still stand as one of the classic pieces of science-fiction writing.

No one would have believed, in the last years of the nineteenth century, that human affairs were being watched from the timeless worlds of space. No one could have dreamed that we were being scrutinized as someone with a microscope studies creatures that swarm and multiply in a drop of water. Few men even considered the pos-sibility of life on other planets. And yet, across the gulf of space, minds immeasurably superior to ours regarded this Earth with envious eyes, and slowly, and surely, they drew their plans against us.

The Mars that H. G. Wells wrote about in 1897 was clearly the Mars described by American astronomer Percival Lowell, who had expanded on Giovanni Virginio Schiapa-relli's initial observations to publish detailed descriptions of Mars as a dying world, whose advanced civilization was desperately trying to keep its desiccated planet alive by drawing water from the Martian ice caps through a system of canals. What would be more natu-ral for such a species than to look down at lush, green and blue Earth and feel envious?

But Mars and the motivations of the Martians were really just a backdrop to Wells's original inspiration for the book—a moment he remembered in detail. He had been walking through the English country-side of Surrey with his brother, Frank, dis-cussing the terrible fate suffered by native populations whenever they were "discov-ered" by European explorers. During the con-versation, his brother said, "Suppose some beings from another planet were to drop out of the sky suddenly and begin laying about them here!"

Instantly Wells saw what his book would be—a cautionary tale of the disaster brought about when so-called advanced societies in-vaded the lands of the less advanced. Only, in his story, the entire Earth would be put in the role of the less advanced society. Mars, with all the public speculation about its possible inhabitants that had been fueled by Lowell's maps and books, was the perfect place of ori-gin for the more advanced society that would invade the Earth, and with the publication of *The War of the Worlds,* literary history was made.

Intriguingly, along with all the other story-telling elements established by *The War of the Worlds,* it is Wells's reminder of what happens when two technologically mis-matched cultures meet that some people point to today as a reason why we should not be too quick to transmit signals into space in an attempt to establish communications with any alien civilizations that might exist on other worlds. If aliens turn out to be more advanced than we are, some people believe our way of life might not survive the contact. For all that Wells looked to the future, per-haps even he would have found it hard to be-lieve that more than one hundred years later, at the dawn of the twenty-first century, his incredible story of alien invasion would still have the power to influence our search for life on other worlds.

us there. Our explorers are all talented and highly trained specialists, scientists, and engineers, but some were also little boys who took apart their mothers' kitchen appliances and struggled to put them back together, while others were little girls who were good in math but never thought of working in space exploration. Still others were jazz musicians, or high school dropouts, and even in the case of Brian,

July 4, 1997, in the Pathfinder Mission Control Room, as signals from Mars confirm the *Pathfinder* lander has successfully touched down on Mars. Blending science with science fiction, Brian Muirhead is the smiling fellow in the middle, with the Jean-Luc Picard haircut. COURTESY NASA/JPL.

a young engineer-to-be who first parlayed his knowledge of motorcycles and domestic cars into becoming a Mercedes-Benz mechanic when he had never opened the hood of a Benz in his life.

For the average reader, this book will tell the story of going to Mars with all the decimal points and the complex scientific exposition that underlie that journey and make it possible, conveniently set off in sidebars—to be read or skipped as desired. (For instance, for those who are interested, this introduction has a "Rocket Science" sidebar that describes the technical details used to define all the different types of years that we mentioned.)

For the fan of science and technology, real and imaginary, other sidebars in this book include the fascinating cultural and decidedly *un*scientific history of Mars, showing how that planet has been portrayed in novels and movies and television. As Carl Sagan noted, science-fiction depictions of Mars can be inspired by science, and as our scientific understanding of the Red Planet has changed over the years, so has our public perception of it.

But for all the different types of information we've included in this book, from discussions of Flash Gordon and Hot Wheels toys to the mathematical complexities of a "pork chop" plot and the fine points of communicating with spacecraft at Mars, there's one thing our story about going to Mars doesn't have—an ending.

It does have a beginning—one that happened thousands of years ago when the first humans noticed that one of the lights in the night sky was a bit redder

than the others and moved on its own. Our story picks up the pace in August 1877, which is when Mars suddenly exploded into the public consciousness as something more than just the fourth planet from the sun. Keep reading and you'll find out why.

Then, like all good thrillers, there's the sudden plot reversal of July 14, 1965, when the first probe to Mars startled scientists with its unexpected findings. We follow that with the enigmatic events of 1976, the nature of which is still being hotly debated today, almost thirty years later. More clear-cut is the triumph of 1997, when a little rover named *Sojourner* first set wheels on the Martian surface (and you've all seen the famous footage of Brian and the *Pathfinder* team cheering and hugging and crying in the mission-control room).

And then, as in any good story, that triumph is followed by the heartbreak of the setbacks in 1999 when two robotic spacecraft were lost at Mars, offering solid proof that more than anything, our journey to Mars is an all-too-human undertaking.

Next, our story takes us into the middle of the excitement and celebration as the people of NASA and JPL pull together like Rocky Balboa getting back on his feet for the final decisive round of the championship match. It's January 2004, time for the most spectacular and successful Mars mission yet—the landing and operation of the Mars Exploration Rovers (MER), *Spirit* and *Opportunity*, even now still functioning as this book goes to press, long months after exceeding their "design life." And you'll find out how the MER team pulled off that incredible success against almost overwhelming odds.

But even then, we're nowhere near the end of our story.

Right now, NASA's Mars Reconnaissance Orbiter is being assembled at Lockheed Martin Astronautics in Denver, Colorado. It's scheduled for launch in August 2005, to photograph details as small as the size of a dinner plate on Mars, mapping new landing sites for the next round of surface explorers.

One of those explorers will be the *Phoenix* Lander, also being built at Lockheed Martin, and scheduled for launch in 2007. (Why launch to Mars every twenty-six months more or less? Check out the "Background" sidebar titled "1877: A Space Odyssey.") In April 2008, *Phoenix* is scheduled to land near the north polar region of Mars and dig three feet into the Martian soil. In those deep trenches, *Phoenix* will use advanced microscopic imagers and sophisticated chemical testing equipment to look for the water ice we believe should be there, and discover what it can tell us about the history of the planet, including the possibility that Martian life-forms—if there were or are such things—might have left some trace of their existence.

Then, there's the proposed Mars Science Laboratory, being designed for launch in October 2009. And as early as 2013, there's even a chance a Mars Sample Return mission will set off to bring back to Earth a little bit of Martian rock and soil (and who knows, maybe even some Martian fossils or, dare we dream, microbes?) for detailed study.

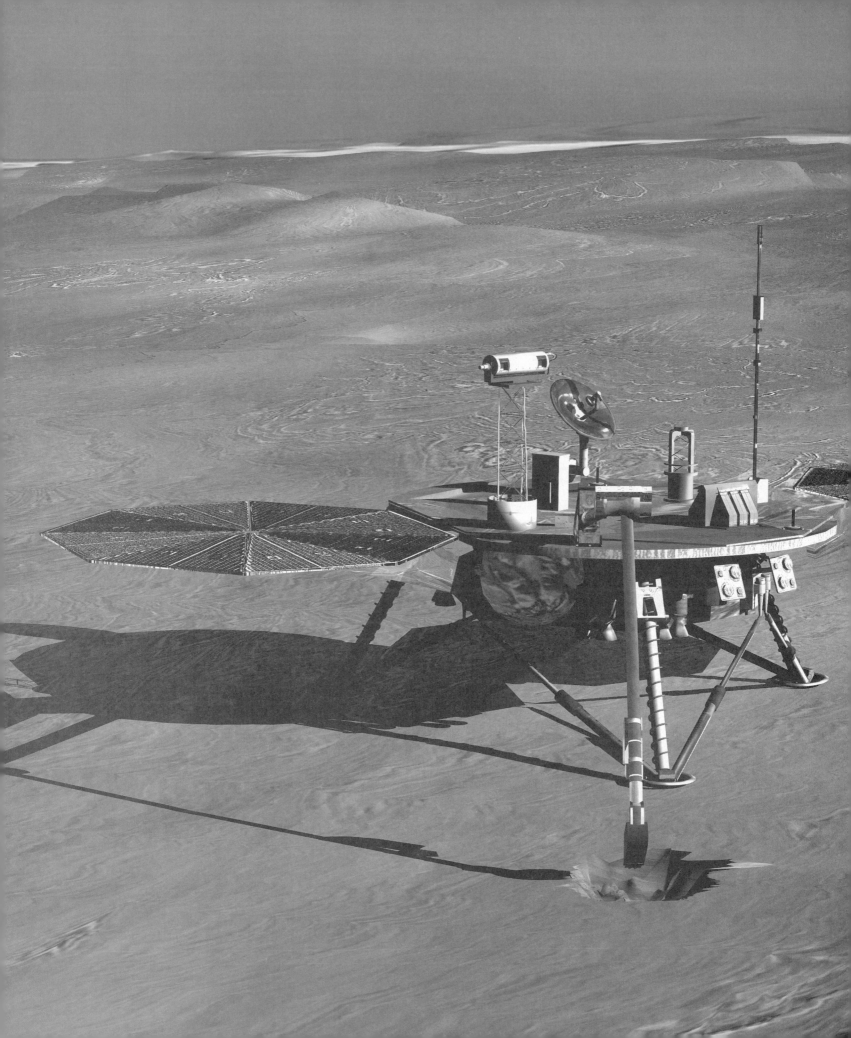

The *Phoenix* lander will set down in the north polar region of Mars in April 2008. Though the spacecraft will not be able to move across the Martian surface, its robotic arm will be able to dig trenches as much as three feet deep to search for water ice and—possibly—detect signs of past biological activity. COURTESY NASA.

# MARS IS FOR EVERYONE

*The Founding Declaration of the Mars Society*

The Mars Society is the single largest private advocacy group specifically dedicated to the exploration and settlement of Mars. It was founded in 1998 at an inaugural meeting of 700 Mars enthusiasts. Today, the Society has thousands of members in eighty chapters around the world, actively involved with private companies and government agencies, including NASA, in a number of significant ongoing studies and simulated Martian scientific outposts.

The Founding Declaration of the Mars Society sets out its rationale for pursuing Mars missions as quickly as possible—ideals that are supported in whole or in part by virtually every similar group, private, public, and governmental—that shares the ultimate goal of going to Mars.

### The time has come for humanity to journey to Mars.

We're ready. Though Mars is distant, we are far better prepared today to send humans to Mars than we were to travel to the Moon at the commencement of the space age. Given the will, we could have our first teams on Mars within a decade.

### The reasons for going to Mars are powerful.

We must go for the knowledge of Mars. Our robotic probes have revealed that Mars was once a warm and wet planet, suitable for hosting life's origin. But did it? A search for fossils on the Martian surface or microbes in groundwater below could provide the answer. If found, they would show that the origin of life is not unique to the Earth, and, by implication, reveal a universe that is filled with life and probably intelligence as well. From the point of view of learning our true place in the universe, this would be the most important scientific enlightenment since Copernicus.

### We must go for the knowledge of Earth.

As we begin the twenty-first century, we have evidence that we are changing the Earth's atmosphere and environment in significant ways. It has become a critical matter for us better to understand all aspects of our environment. In this project, comparative planetology is a very powerful tool, a fact already shown by the role Venusian atmospheric studies played in our discovery of the potential threat of global warming by greenhouse gases. Mars, the planet most like Earth, will have even more to teach us about our home world. The knowledge we gain could be key to our survival.

### We must go for the challenge.

Civilizations, like people, thrive on challenge and decay without it. The time is past for human societies to use war as a driving stress for technological progress. As the world moves towards unity, we must join together, not in mutual passivity, but in common enterprise, facing outward to embrace a greater and nobler challenge than that which we previously posed to each other. Pioneering Mars will provide such a challenge. Furthermore, a cooperative international exploration of Mars would serve as an example of how the same joint-action could work on Earth in other ventures.

### We must go for the youth.

The spirit of youth demands adventure. A humans-to-Mars program would challenge young people everywhere to develop their minds to participate in the pioneering of a new world. If a Mars program were to inspire just a single extra percent of today's youth to scientific educations, the net result would be tens of millions more scientists, engineers, inventors, medical researchers, and doctors. These people will make innovations that create new industries, find new medical cures, increase income, and benefit the world in innumerable ways to provide a return that will utterly dwarf the expenditures of the Mars program.

### We must go for the opportunity.

The settling of the Martian New World is an opportunity for a noble experiment in which humanity has another chance to shed old baggage and begin the world anew; carrying forward as much of the best of our heritage as possible and leaving the worst behind. Such chances do not come often, and are not to be disdained lightly.

### We must go for our humanity.

Human beings are more than merely another kind of animal—we are life's messenger. Alone of the creatures of the Earth, we have the ability to continue the work of creation by bringing life to Mars, and Mars to life. In doing so, we shall make a profound statement as to the precious worth of the human race and every member of it.

### We must go for the future.

Mars is not just a scientific curiosity; it is a world with a surface area equal to all the continents of Earth combined, possessing all the elements that are needed to support not only life, but technological society. It is a New World, filled with history waiting to be made by a new and youthful branch of human civilization that is waiting to be born. We must go to Mars to make that potential a reality. We must go, not for us, but for a people who are yet to be. We must do it for the Martians.

Believing therefore that the exploration and settlement of Mars is one of the greatest human endeavors possible in our time, we have gathered to found this Mars Society, understanding that even the best ideas for human action are never inevitable, but must be planned, advocated, and achieved by hard work. We call upon all other individuals and organizations of like-minded people to join with us in furthering this great enterprise. No nobler cause has ever been. We shall not rest until it succeeds.

Past that? Well, the sky truly is the limit. With every mission come new discoveries that will drive the missions after. There will be more orbiters with specialized sensing equipment, more long-duration rovers, even the possibility of a race of sorts with our friends at the European Space Agency looking for definitive proof of whether Mars has ever harbored life.

One day, inevitably and inescapably, there will come launches of components that will form the infrastructure of the first human outpost on the Red Planet. And then, finally, with great fanfare, there will be another launch—the first humans bound for Mars.

When that historic mission will launch, and who will be part of it, are details still to be determined, which is why our book and its real-life story have no ending. In fact, considering that the first human expedition to Mars will someday be followed by the second, and that, in an even more distant someday, the people who travel there won't be explorers but colonists . . . no book about Mars can ever have an ending, because there will always be more of the story waiting to unfold.

But this book is the beginning.

And if the three of us can make the first of many predictions that are in these pages, the one thing we can say about the first explorers to go to Mars, whoever they may be, is that they will be scientists and engineers who live and work in a world of science—of numbers and precision—and who were inspired by *Star Trek* and *Babylon 5*, by Arthur C. Clarke, Ray Bradbury, and Kim Stanley Robinson . . . by science fiction.

And, we hope, perhaps even by this book.

—BRIAN MUIRHEAD
JUDITH & GARFIELD REEVES-STEVENS

# CHAPTER ONE

# "Six Minutes of Terror"

## Entry, Descent, and Landing—January 2004

*They told me, What we're going to do, very, very simply, is take this thing onto*

*Mars, throw this big entry-vehicle bullet in, a parachute is going to open up,*

*this lander's going to rappel down this long rope, and some airbags are going to*

*inflate, and it's going to hit the ground, bounce, and open up like a flower, and*

*this rover's going to drive off.*

*And I go, Really?*

*I thought they were nuts, actually.*

—ROB MANNING, LEADER, ENTRY, DESCENT, AND LANDING DEVELOPMENT TEAM
MARS *PATHFINDER* AND MARS EXPLORATION ROVERS

## EARTH STRIKES BACK

Martians first invaded Earth in 1898. They lost, soundly defeated by Earth germs, as reported by British writer H. G. Wells in his novel *The War of the Worlds.*

Forty years later, the persistent Martians tried again, this time on the night of October 30, 1938. From a ringside seat to this sinister invasion attempt, the brilliant young actor/director Orson Welles and his Mercury Theatre Players delivered an emotional, blow-by-blow account by radio. Though some of the American radio audience were unnerved, once again good old Earth germs stopped the Martian war machines in their tracks.

For the next twenty-seven years, those pesky Martians kept trying. Countless times they landed on movie screens around the world, inserting wriggling bugs into the backs of our heads to control our thoughts, claiming they needed our women, and otherwise causing a great deal of trouble for us puny Earthlings. But only for twenty-seven years.

Because, in 1965, we humans struck back.

With solar panels open, the *Mariner 4* probe measured twenty-two and a half feet wide. Though its nineteen useable images of Mars covered only 1 percent of the planet's surface, the detail they provided in 1965 forever changed the way we thought about the Red Planet. COURTESY NASA.

True, our technology wasn't sophisticated. Back then, only a handful of astronauts and cosmonauts had orbited the Earth. Only a few robotic craft had landed on the Moon, and a few others that had headed there had actually *missed* their target!

But for all the decades and centuries that Mars had tempted us, for all the invasions the Martians had sent our way in science-fiction novels and movies, *we* finally had the ability for the first time in history to send a device of our own making to really look at Mars, up close and personal.

The spacecraft that made this historic journey—the first from Earth to go to Mars—was *Mariner 4*, a 575-pound collection of scientific instruments and solar panels with less onboard computing power than a present-day calculator watch. Like many of the best planetary probes to follow, it was built by the Jet Propulsion Laboratory in Pasadena, California, otherwise known as JPL.

As befits the *Mariner*'s pioneer status, full facilities to support planetary exploration didn't yet exist at JPL, so some of the final assembly of the craft took place in a JPL parking lot, with clean-room conditions provided by a jury-rigged structure of scaffolding and plastic sheeting.

As for why the spacecraft was called *Mariner*, that name came from what NASA termed "the Cortwright system" of naming space probes. In 1960 Edgar Cortwright was the space agency's director of lunar and planetary programs, and he proposed

# MARTIANS INVADE NEW JERSEY!

*Did Anyone Really Notice?*

Most everyone has heard of "The Night that Panicked America"—the notorious 1938 Halloween radio broadcast of *The War of the Worlds* that caused mass hysteria across the United States as millions of people convinced the Earth was being invaded by Martians, fled their homes.

Then again, most everyone's heard of the canals of Mars, too, and they don't exist any more than that off-cited night of hysteria took place as legend describes it.

The truth is, some radio listeners did panic, becoming confused and upset by Orson Welles's clever and dramatic updating of the classic H. G. Wells story. At the time, the population of the United States was about 130 million. A sociological study of the broadcast's effect, published in 1940 by Hadley Cantril of Princeton University, concluded that of the approximately 1.7 million people who tuned in for at least part of the broadcast, more than a million were "severely alarmed" by what they heard. Though subsequent studies have drastically reduced that estimate, it is intriguing to look into the events of that night to find out what kind of hold the idea of Mars and Martians had on Americans almost seventy years ago.

First, though, we must acknowledge that factors other than Mars played a part in the reaction of at least some of the public. Just as H. G. Wells was the first to write a modern alien invasion story, Orson Welles was the first to broadcast a mock documentary. Considering that a real war was underway in Europe (complete with news bulletins interrupting regular radio broadcasts to report military action overseas), and that many Americans were already concerned that the conflict might spread to the United States, the reaction to the radio play is generally considered to be more a result of "war jitters" than a genuine fear of alien invasion.

Certainly, many more listeners recognized Orson Welles's distinctive voice as that of the popular radio character the Shadow than believed they were listening to an oddly time-compressed series of real, as-it-happens news reports.

Of course, given what was known about Mars in 1938, the existence of a technologically advanced civilization on that planet could not completely be ruled out. The question of Percival Lowell's canals was still up in the air, and many astronomers accepted that the seasonal color variations they saw on Mars might be caused by vegetation.

But just as we know considerably more about Mars today, we also know more about what happened on the night of October 30, 1938. The *New York Times* for Monday, October 31, 1938, carried the front-page headline: "Radio Listeners in Panic, Taking War Drama as Fact: Many Flee Homes to Escape 'Gas Raid from Mars'—Phone Calls Swamp Police at Broadcast of Wells Fantasy."

The news article goes on to describe the country's reaction to the broadcast, including descriptions of church services being interrupted and restaurants closing, while hundreds of people on the East Coast left their homes to avoid the advancing Martian hordes, as those on the West Coast sought to join the fight. In New Jersey, ground zero for the Martian beachhead, hundreds of doctors and nurses called local hospitals to volunteer for emergency service, and two geologists from Princeton University loaded their car with scientific equipment to search for the meteorite the radio said had killed fifteen hundred people when it hit the ground. Other people near the site of the supposed Martian landing phoned police and their friends to report they could see the smoke and flames from the first attacks.

But interspersed with these sensational vignettes of public reaction, the article also explained that *The War of the Worlds* was listed in the newspapers' radio guides for that night, and that five times during the hour-long broadcast of the regularly scheduled *Mercury Theatre on the Air,* the evening's entertainment was clearly identified as a radio play. The article also quotes the AP newswire announcement that reminded news departments across the country that the events in New Jersey were imaginary, as well as the bulletin issued by the New York Police Department informing their officers that there was no reason for concern.

In fact, once the sensationalism is stripped from the breathless reports of the time, and each story of a switchboard being overwhelmed by calls (not a difficult challenge on a Sunday night in 1938) is matched with a story of someone's indignation at having been confused by tuning into the broadcast a few minutes late and not paying attention to the announcer, it's easy to see that while some people panicked that night, when it comes to the vast majority of the population, America didn't.

Of course, that's not to say that the power of Mars on human imagination can entirely be discounted. Since its original 1938 performance, at least four other radio versions of *The War of the Worlds* have also instilled panic in listeners. On November 12, 1944, the version broadcast from a radio station in Santiago, Chile, resulted in a provincial governor mobilizing troops to fight the Martian invaders.

A different version of the story, broadcast on February 12, 1949, resulted in thousands of residents of Quito, Ecuador, abandoning their homes to escape a Martian gas attack. That panic turned tragic when the people of Quito realized the news reports were a fictional presentation, and an angry crowd burned down the radio station responsible, resulting in fifteen deaths.

And as recently as 1974 in Providence, Rhode Island, and 1988 in Portugal, new adaptations of the play caused more instances of "severe alarm" among listeners.

When it comes to telling and retelling the stories of *The War of the Worlds* radio broadcast, once again people look to Mars and just like Percival Lowell, they often see what they expect to see, and not what's really there.

## ONE-WAY LIGHT TIME

*How Far Away Is Mars, Anyway?*

Neither Earth nor Mars orbits the Sun in a perfect circle. Instead, each of their orbits is an ellipse, which means that each planet's distance from the Sun varies throughout the course of its journey around it, sometimes closer, sometimes farther away (although for most practical purposes Earth's orbit is nearly circular).

For convenience, astronomers average out these orbital variations to arrive at what they call a planet's "mean" distance from the Sun. For example, over the course of the year, the Earth moves as close to the Sun as 91.5 million miles, and as far away as 94.5 million miles. However, as a useful shorthand, the Earth's mean distance from the Sun is set at 93 million miles, which is also called 1 AU, for "astronomical unit." For Mars, which varies between 128 and 154.1 million miles from the Sun, the mean distance is given as 141.5 million miles (or 1.5 AU).

Because Earth is closer to the Sun than Mars is, we travel around the Sun more quickly—in 365.25 days to be exact, which is the length of an Earth year. Mars, on the other hand, takes 687 Earth days to complete

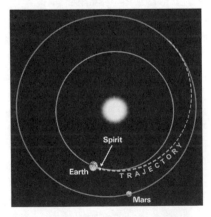

As Earth and Mars follow their separate orbits around the Sun, the distance between them constantly changes, as does the time it takes for radio signals to travel from one to the other. COURTESY NASA.

that lunar probes be named after land exploration activities—hence *Ranger* and *Surveyor*—and that planetary probes reflect nautical terms to express "the impression of travel to great distances and remote lands." Thus, the Cortwright system also established the names for the *Mariner* and *Voyager* probes, as well as for the *Viking* missions to Mars. (The tradition of calling robotic spacecraft "probes" dates back to a paper presented to the British Interplanetary Society in 1952. The paper was titled "The Martian Probe," and the term has stayed with us ever since.)

As for why this first mission to Mars was given the number 4, there's good news and bad news. The first two *Mariners* were Venus probes. *Mariner 1* was launched on July 22, 1962, but had to be deliberately destroyed when it went off course five minutes after liftoff. *Mariner 2* launched a month later on August 27, and after a voyage of three and half months, became the first probe to successfully fly by Venus and return scientific information. (The Russian probe *Venera 1* had actually flown by Venus in 1961 as planned, but since all communications were lost prior to its arrival, the mission is not considered a success.)

The next two *Mariners*, *3* and *4*, were Mars probes. *Mariner 3* suffered a fatal malfunction when its protective shroud failed to separate from the spacecraft after launch, so it was up to *Mariner 4* to make history.

And it did.

Up until July 15, 1965, Mars was a blurry object that human eyes had never seen except through the thick, churning blanket of Earth's atmosphere. The same phenomenon that makes stars appear to "twinkle" makes everything in space observed from the Earth's surface seem to shift and waver. Think of the distorted view you see when you peer through a column of heated air rising above a radiator or a paved road in the summertime, and you'll get an idea of what Earth-based astronomers have to deal with every night. Atmospheric distortion is why observatories are so often located on the highest mountaintops possible—the higher the telescope, the less atmosphere there is to look through.

Distance also plays a part in what we can see in space. The Hubble Space Telescope, which is rightly considered one of NASA's greatest success stories, has amazed and astounded the general public *and* jaded astronomers for more than a decade with its spectacular images of objects in space, taken well outside the obscuring effects of Earth's atmosphere. But at distances that range from 33.5 to 250 million miles from the Earth to Mars, even the Hubble's outstanding optics can't compete when it comes to capturing images of the Red Planet from only a few thousand miles away.

Of course, back in 1965, there was no Hubble Space Telescope. It was launched in 1990 and became fully operational following a space shuttle mission to correct its faulty main mirror in 1993. So the images captured by the black-and-white television camera system on *Mariner 4*, from a distance of 6,188 miles above the planet at its closest approach, were widely anticipated to reveal surface details no one had ever seen before.

Unless, of course, those canals turned out to be real.

Today, we live in a world of instant gratification. When a space probe goes into orbit of another planet, we see pictures right away. In the summer of 2004, months before it even entered orbit of Saturn, NASA's successful *Cassini* probe began transmitting images of the ringed planet and its moons.

But in 1965, patience was not only a virtue, it was a necessity. The *Mariner 4* probe had taken seven and a half months to reach Mars. During its flyby, over the space of 21 minutes, 36 seconds, *Mariner 4* took eighteen useful pictures (and four and a bit unusable ones) which its onboard system recorded on 330 feet of magnetic tape in a forerunner of videotape technology. (The first image captured by *Mariner 4* showed no discernible features, and the last three plus a final partial image were of the planet's nightside, and again the camera system could capture no detail.)

By present-day standards, which allow the latest Mars Exploration Rovers to return full-color, three-dimensional images of Mars with enough detail to be projected on IMAX movie screens seventy feet tall, the first images returned from *Mariner 4* were decidedly primitive.

Each image consisted of only 200 scan lines, each with 200 pixels of information, each 6-bit pixel capable of representing one of 64 different shades of gray.

---

For the technically minded, that works out to 240,000 bits of information per image. For comparison, ordinary television sets in North America—then and now—display images made up of 485 scan lines. Some current high-definition televisions can display 1,080 lines, and each of those lines can hold up to 1,920 pixels of information, with each 24-bit pixel capable of representing any one of *millions* of colors. Compared to the 240,000 bits making up each *Mariner 4* image, a single frame of a high-definition television signal contains almost *50 million* bits of information.

---

The total information content of the *Mariner 4* images, though, was only part of the story. Each one, as limited as it was, had to be recorded on magnetic tape for later playback because, at a maximum transmission rate of 33.3 bits per second, it took *eighteen days* for all the images to be transmitted back to Earth. If you'd like another point of comparison, transmitting the same amount of information by ordinary dial-up modem for a home computer today would take less than five minutes, and that includes sending each image twice so transmission errors can be detected. (And by high-definition television broadcast standards, that same amount of information could be transmitted in about a quarter of a second.)

So, the first Earth object to go to Mars was a spacecraft built in a parking lot that sent back images less than half as detailed as those that appeared on a 1965 black-and-white television set. And, as an added kicker, those images covered only about 1 percent of the surface of Mars, and no one knew *which* 1 percent.

Was it worth it?

In a word, absolutely.

Because those nineteen usable pictures, and the results from the *Mariner*'s other scientific instruments, changed forever how we thought about Mars.

First, and most dramatically—because no one was expecting it—the images showed craters, lots of craters.

its orbit, so that's the length of the Martian year.

Because the Earth travels around the Sun faster than Mars, half the time it's gaining on Mars, like a race car speeding around an inside curve. The other half of the time, it's pulling away from Mars. So just as the Earth's distance from the Sun changes throughout the year, so does its distance from Mars.

Whenever Earth catches up to Mars—that is, gets as close as it can, so that the Sun, Earth, and Mars are more or less aligned—astronomers say the two planets are in "opposition." That's because from the perspective of an observer on Earth, Mars and the Sun appear in directly opposite parts of the sky. On average, this happens about every 780 Earth days, which is also the average time between NASA's launch windows for sending spacecraft to Mars. Overall, though, oppositions can occur as close together as 761 days, or as far apart as 805 days.

Sometimes, about every 284 years or so (because of the elliptical shape of the orbits), opposition occurs when Earth is at its closest to the Sun, and Mars is at its most distant. At that time, the closest the two planets can get is 62.6 million miles. Other times, opposition occurs when Earth is the farthest it gets from the Sun and Mars is the closest. At those times, the two planets can get to within 33.5 million miles. The closer the planets get, the better the opposition is considered to be because the easier it is to observe Mars and to send spacecraft to it. In the opposition of August 27, 2003, Mars came within 34,646,418 miles of Earth, the closest it had been in more than 59,000 years.

Part of the reason for why some oppositions are closer than others is that not only are Mars and Earth traveling in different-shaped orbits, the plane of their orbits also vary.

One way to picture this arrangement is to imagine the Earth's orbit around the sun drawn on the flat surface of a table. Because the orbit of Mars is tilted in respect to the Earth's orbit, sometimes Mars is following its orbital path above the tabletop, and sometimes it's on a path below the tabletop. So, in addition to calculating the distance between Earth and Mars in terms of how far away each is from the Sun, it's necessary to calculate how far "above" or "below" the Earth Mars happens to be.

For the opposition of August 2003, at the same moment Earth and Mars were at their closest approach in terms of their distances from the Sun, the position of Mars in its orbit was very near the orbital plane of Earth as well.

Knowing the constantly varying distance between the Earth and Mars is critical for calculating the path spacecraft must take to travel from one planet to another. After all, a spacecraft launched from Earth can't be aimed at where Mars *is* at the time of launch, but where it *will* be six to twelve months later.

The distance between the planets is also a crucial factor in determining how and when communication can take place between Mars spacecraft and Earth control centers. Space mission planners use the phrase "one-way light time" (OWLT for short) to describe the length of time it takes radio signals—traveling at the speed of light—to leave the Earth and reach Mars, or to leave Mars and reach the Earth. That interval varies according to distance, so that when Mars and Earth are at their closest, the one-way light time can be as short as three minutes, and when they are at their most distant, as long as twenty-two and a half minutes.

Technically, though, when Mars and Earth are as far apart as possible in their orbits, they're on opposite sides of the Sun, in what's known as solar conjunction, so direct communication may not be possible for days to weeks.

NASA's Hubble Space Telescope provides astronomers with the best images of Mars available from Earth. But even at the Red Planet's closest approach, the smallest surface features Hubble can reveal are twelve miles across. At its closest approach, the *Mariner 4* probe was able to resolve features about nine-tenths of a mile across. Using special techniques, today's Mars Global Surveyor has been able to reveal details as small as five feet across, enabling it to detect the Mars *Pathfinder* and Mars Exploration Rover landers. At its highest resolution, the Mars Reconnaissance Orbiter—scheduled for launch in August 2005—will be able to detect details as small as ten inches. COURTESY NASA.

Now, every planet in the solar system has been bombarded with meteoroids and comets since the planets first formed, about 4.6 billion years ago. Since the Earth and the Moon are basically in the same location in the solar system, scientists assume that both have been subjected to the same intensity of bombardment for those billions of years. Yet, for all the craters that we see on the surface of the Moon, very few are easily visible on the surface of Earth. That's because Earth has two things the Moon doesn't. First, the Earth has a dynamic atmosphere that produces energetic rain and wind and glaciers and floods to erode craters. And, second, Earth has plate tectonics, which produces volcanoes and lava fields and actually recycles the outer surface of our planet by pushing it up in some regions to form mountains, while drawing it back down in other regions to melt deep below the crust. (And yes, science fans, we are simplifying.)

So, the discovery of craters on Mars in Moonlike abundance instantly suggested to scientists that Mars had neither a dynamic atmosphere nor a dynamic crust. We say "suggested" because scientists do not like to jump to conclusions based on limited information, especially when it's the first information they get. Interestingly enough, though, that 1 percent of the Martian surface that *Mariner 4* imaged *did* turn out to be, just by random chance, in the most heavily cratered part of the planet. Other regions have almost no craters, mostly in the northern hemisphere, because lava flows appear to have covered over the earliest ones.

Without a "dynamic" atmosphere, the case for complex life on Mars took a big hit. For decades, astronomers had charted seasonal variations in the colors of the Martian surface that corresponded to the planet's changing seasons. Based solely on that visual information, it was reasonable to propose—though by no means definite—that the color changes might be caused by fields of vegetation coming to life in the Martian spring and dying off in the Martian winter.

But if the atmosphere of Mars was too thin to produce weather that would erode craters over millions of years, how could it hold enough moisture for vegetation to

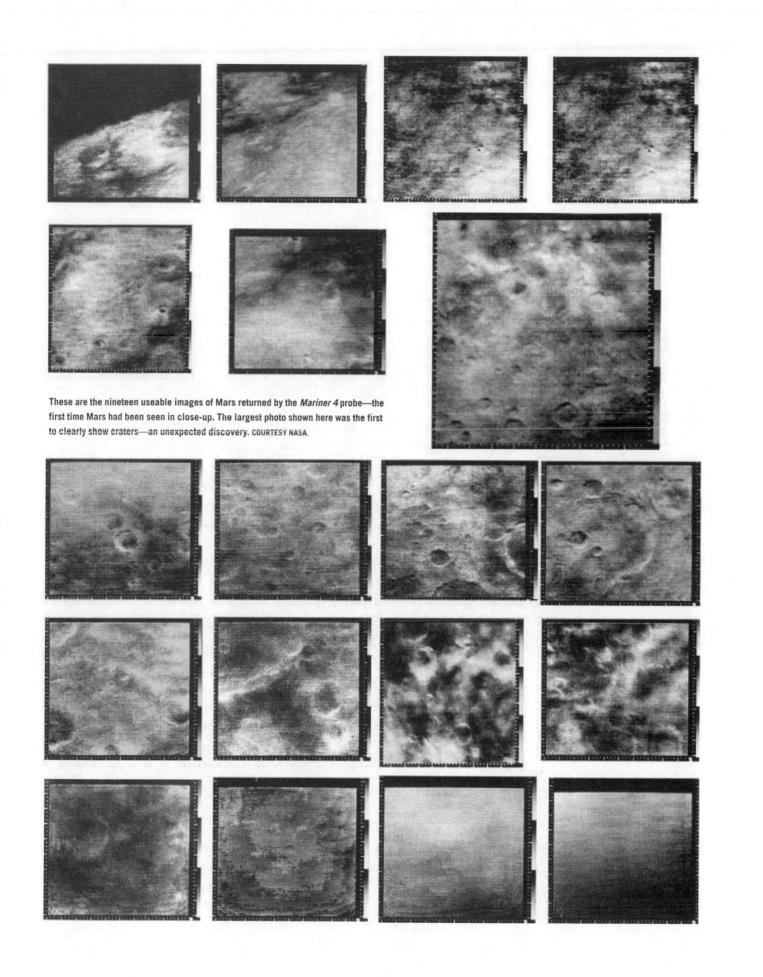

These are the nineteen useable images of Mars returned by the *Mariner 4* probe—the first time Mars had been seen in close-up. The largest photo shown here was the first to clearly show craters—an unexpected discovery. COURTESY NASA.

survive? How could it hold enough oxygen to support any Martian animals that might eat that vegetation?

And where were those canals, anyway?

*Mariner 4* provided even more information about the Martian atmosphere when its course took it behind the planet. For a few brief moments, the radio signals it sent back to Earth were transmitted through the Martian atmosphere. By measuring the changes in that signal, scientists were able to calculate the atmosphere's thickness.

Those calculations agreed with the visual evidence of the craters.

---

According to the radio signal, the Martian atmosphere was estimated to have a pressure of between 4.1 to 7.0 millibars. At sea level on Earth, average atmospheric pressure is considered to be 1,013 millibars. In another useful comparison, on the peak of Mount Everest, about 5.5 miles above sea level, the average atmospheric pressure is about 300 millibars. Since those 1,013 millibars of pressure at sea level convert to around 14.7 pounds of pressure per square inch, that makes the air pressure on the peak of Mount Everest about 5 pounds per square inch.

---

Those first measurements right away told scientists that the atmosphere of Mars was less than 1 percent as thick as Earth's. Any human trying to breathe that atmosphere would black out within seconds and die as quickly as if he were in the vacuum of space. Overnight, the idea that the Red Planet was home to a race of intelligent Martians moved decisively and permanently back to the realm of science fiction.

But in the realm of real science, now that scientists had real Martian data to study, plans for Earth's next assault on Mars could begin in earnest.


## THE DR. DOLITTLE CONNECTION

NASA's first plans for planetary exploration were modeled on the approach that was working so well in the effort to eventually land astronauts on the Moon: first, have a probe fly by the target planet; next, have a probe go into orbit around it for closer observation; and third, land robotic craft.

Almost twenty-nine years after *Mariner 4*, that probe's first measurements of the Martian atmosphere, refined over the many years and missions that followed, were again put to the test as Earth began its most ambitious assault on Mars yet— landing two robotic rovers on the surface within a month of each other.

One percent of Earth's atmosphere might not sound like much. But to a spacecraft traveling at a speed of about 12,000 mph, even the Martian atmosphere is thick enough to mean instant and total destruction if anything at all goes wrong.

On the night of January 3, 2004, a new generation of scientists and engineers gathered at the Jet Propulsion Laboratory in Pasadena, California, all of them hoping for success, yet bracing for failure. Some had parked their cars near where *Mariner 4* had been assembled so many years earlier. But this time, the spacecraft they were following had been built indoors, in the Lab's "High Bay One"—an immense, stark-white clean room in the Spacecraft Assembly Facility, with a forty-five-foot-tall ceiling.

After three years of intense effort, *Spirit*, the first of two Mars Exploration Rovers, had reached the final six minutes of its journey to Mars—six minutes in which its speed had to drop from 12,000 to 0 mph. The craft's success was completely dependent on more than sixty-six complex mechanisms working perfectly after almost seven months in the vacuum of space; the same mechanisms that the designers had protected, so they hoped, from the brutal cold of space during cruise, and now from external temperatures in excess of 2,900°F.

The technical name for this portion of the mission is EDL, for Entry, Descent, and Landing. But most people on the EDL team think of it as "six minutes of terror," because all they could do was wait and listen for signals from the spacecraft that it was doing what they designed it to do and dealing with the uncertainties they hoped they had anticipated, each of them knowing that at any instant transmission could cease forever because something had gone terribly wrong, wiping out years of labor and dreams in a heartbeat.

Some members of the Mars Exploration Rover team were new to the experience of EDL. But for a special core group of veteran scientists and engineers, the events of this night weren't just a stage of going to Mars for the first time—it was their *return* to the Red Planet.

Six and a half years earlier, this core group had gathered in the same building at JPL, sweated through an almost identical EDL sequence with the rest of their team, waiting to find out if their audacious new approach to planetary exploration was going to pay off with *Pathfinder*'s landing of the first rover on Mars.

And for all the science and engineering, all the innovation, hard work, sleepless nights, missed birthdays, anniversaries, and vacations that led up to those anxious but thrilling minutes of EDL in 1997 and 2004, the real reason all these explorers had gathered together at JPL on both occasions, was because fifty years earlier, Dr. Dolittle had gone to the Moon.

You're right, we should probably explain that.

The *technical* path to Mars is pretty straightforward. We start at Earth, pick the days we want to leave and arrive that will give us the "lowest-energy trajectory" (which means it will require the smallest, least-expensive rocket that can lift our spacecraft), and these days, with the punch of a button, a wonderfully complex piece of computer programming developed at JPL will spit out the smoothly curving trajectory that our craft will follow.

But what about the *people* who create the program by which we go to Mars? What about the people who figure out how to slow a spacecraft down from 12,000 to 0 mph in six minutes? The people who hand build the computer boards that will let a rover decide how to avoid a rock on its own, instead of waiting for commands from Earth. The people who not only dream about going to Mars, but who have put themselves in a position where they can actually do something about it.

In other words, what is the *human* trajectory for going to Mars?

## JOHN CARTER OF MARS

*One Giant Leap of Imagination*

In 1911, a pencil-sharpener salesman in Chicago struggling to support his wife and two children, began scribbling a story on the back of sheets of used letterhead. The salesman's name was Edgar Rice Burroughs, and he was destined to become one of the most successful popular writers of all time because of his creation, *Tarzan of the Apes.* But in 1911, Tarzan was only a half-formed idea in the back of Burroughs's mind. The story he wrote first was *A Princess of Mars. Princess* was serialized in *Argosy* magazine from February to April 1912, and Mars was never the same again.

*A Princess of Mars* told the story of Captain John Carter, a Civil War veteran who one day escapes his pursuers by hiding in a cave. For no reason at all—by the storytelling conventions of 1911, none was required—Carter falls asleep, has an out-of-body experience, and gazing up at the shimmering red dot of the planet Mars, is suddenly transported there.

What follows next sets the stage for an ongoing series of incredible adventures on a Mars deeply influenced by the writings of Percival Lowell. Burroughs's Mars—or Barsoom, as its inhabitants call it—is a dying planet scarred by dried-up seabeds, a thinning atmosphere maintained by factories, and home to an ancient civilization. It's also home to a variety of life-forms, including Tharks—green-skinned, four-armed, fifteen-foot tall warriors—and, of course, the Martian Princess, Dejah Thoris, resident of the city of Helium, who appears perfectly human, except for her reddish copper skin and the fact that her species reproduces by laying eggs. Her species also seemed to disdain the wearing of clothes of any kind, no doubt adding immensely

The quick answer is that there isn't just one path, there are an infinite number, each as individual as the people who create and follow them.

But for each Mars mission run by JPL or another NASA facility, the one thing shared by everyone in the flight control room, the one common experience in their unique and personal journeys to those six minutes of entry, descent, and landing on another planet and the adventure that follows, is that at some point in their lives, they experienced the *spark*.

That spark arrives in different ways for different people. For some, it comes so early in life, the person has trouble remembering a time she or he *wasn't* drawn to the exploration of space. For others, it comes so late, that it takes conscious and determined effort to refocus their life and reenter school to begin the accumulation of knowledge they will need to join the spacefaring community.

And for others, it's a more gradual process, made easier because the spark is something that's shared by a parent who always wanted to or does work in a field related to space exploration.

Even if it's only realized in hindsight, if they take the time to reflect on how they got here from there, almost all who end up exploring space *can* pinpoint the time when the spark hit them. For most explorers, there's a sudden, illuminating moment when the clouds part and it becomes clear what they want to do with their lives. That's when they take the first step on their path.

Here's the story of one of those paths taken by someone who, perhaps most significantly, has led NASA and the world to our current approach to go to Mars. We'll find his story, and the moment of his spark, by looking back—way back, to the children's book, *Dr. Dolittle in the Moon.*

As the world knows, and as this book will detail in the chapters ahead, the Mars Exploration Rover missions of 2004 have been an unqualified, spectacular success—NASA and JPL and an international aerospace community of scientists and engineers at their best.

There are egos among them, to be sure. There are rivalries and competitions. At times of stress, there can be sharp disagreements and dissension. The scientists and engineers of JPL and NASA, and the institutions that they are a part of, aren't robotic paragons of perfection, no matter how perfect their work must be. They're still human, and their interaction with one another as in any other workplace still reflects all the conflict and complexity that being human entails.

Thousands of people contributed to the amazing achievements in the 2004 missions to Mars. Hundreds of people made critical decisions and outstanding contributions that paid off in scientific discoveries and technological breakthroughs that will inform the next missions to Mars for years to come. Crucial to those decisions and contributions was a smaller group of people, responsible for the previous great landing success on Mars—the men and women of the 1997 Mars Pathfinder mission.

And for all the talent and expertise that made the Pathfinder mission a breakthrough success against almost impossible odds, one person stands out as the "father" of the mission—the person who, at the very beginning, when a single word could have canceled the mission before it began, said, instead, "Let's do this."

The father of *Pathfinder* is Dr. Wesley T. Huntress Jr., and to describe him is to describe almost every other scientist and engineer long involved in NASA's exploration of space. Because NASA's accomplishments far outweigh the failures that are an inevitable consequence of pursuing cutting-edge, first-of-a-kind, exploration, what most of these explorers share is an inner sense of satisfaction, a deep and deserved sense of accomplishment that arises from having faced daunting odds and impossible tasks and prevailed, often coupled with the drive to take on the next Mission: Impossible.

Wes Huntress is no exception.

In his case, his drive for accomplishment in space exploration—the moment he experienced the *spark*—began in the late 1940s when his first-grade teacher read to her class the 1928 novel *Doctor Dolittle in the Moon.*

It should come as no surprise that there is not a lot of science in this children's storybook. Dolittle and his friends travel to the far side of the Moon on a giant lunar moth, and discover a community of intelligent plants—Moon lilies to be exact. The Moon's lack of atmosphere isn't addressed, though the good doctor enjoys bounding over the lunar surface in the Moon's reduced gravity.

But for a youngster who already had come to realize he liked to know how things worked, who devoured comic books and built models of planes, Wes Huntress recalls, "That was one that got me started on space, got me thinking, What's it like on the Moon? Can you breathe on the Moon? Now I know you can!"

At the same time his first-grade teacher started him thinking about the Moon, young Wes had always had a fascination for flying machines. But that fascination didn't exactly run in the family. Though Wes was the first on both sides of his family to go to university, his father was an aeronautical engineer for the Civil Aviation Authority, forerunner of today's FAA, the Federal Aviation Administration. "You can't do that anymore," Wes says about his father's career as an engineer without a degree, "but he joined the CAA in 1926 and he worked for them until he retired."

His son points out that Wes Huntress Sr.'s area of expertise was airport safety, not aircraft. "He was scared to death of airplanes. He would never get on one. The reason was because he saw too many of his friends get killed in those early days. He didn't want to risk it."

So young Wes's interest in airplanes was indulged by his grandfather. "One of my favorite things about staying with my grandfather on weekends," Wes remembers, "was that he would take me down to Washington National Airport to watch the planes take off and land. I'd go outside on the balcony and watch the planes pull up, see the people get on them, and I wanted to go on one of those things."

Wes's grandfather supplemented those trips to the airport with books about planes and about space travel—still a far-out topic in the late '40s, better suited to Saturday afternoon serials like *The Purple Monster Strikes* (yet another Martian invasion) and *King of the Rocket Men.*

But that was all right with Wes. "I read science fiction like crazy. It started with comic books, and then I started reading science fiction when I was about in sixth grade."

It was also in sixth grade that Wes discovered science and how much he liked it.

to the story's appeal for *Argosy*'s young male readers.

Burroughs's ongoing tales of John Carter were fantastic amalgams of swordplay, intrigue, and derring-do on a Mars that could exist only in fantasy. Even as the popularity of Burroughs and his creations soared, the category of story he wrote, called "scientific romance," gave rise to a new genre known as "scientifiction."

As the pace of technological development and scientific discovery quickened in the first half of the twentieth century, readers would soon need a better explanation for how Earthlings traveled to Mars and more believable descriptions of conditions there to match the theories of the day. New generations of science-fiction writers were in the wings, ready to do exactly that.

Once again, Mars was letting its influence be felt in the affairs of Earth.

Then, when he was a sophomore in high school, he experienced a second spark that was shared by the nation.

The date was October 4, 1957, and today, almost fifty years later, Wes clearly recalls what happened. "I remember waking up one morning and walking outside to go to the bus stop to go to school and the paper was on the doorstep and it was right there—*Sputnik*."

The Soviet Union had jump-started the space age with the launch of the first artificial satellite—a 184-pound orbiter that did little more than transmit a radio beep to allow it to be tracked, and to let the United States know that as *Sputnik* passed overhead every ninety minutes, the Russians were winning the space race.

By that time, Wes already knew he wanted to be a chemist. Thinking back on those early days, he laughs as he remembers, "I got a chemistry set when I was eleven, and I had a great time mixing stuff up. I just loved mixing things. Stuff happened! You mix two things together and they turn blue! That's cool!"

But it was *Sputnik* that gave his interest in chemistry a focus. "That's when I got interested in rocket fuels and building rockets." Thousands of other high school students shared Wes's sudden interest in rocketry. Unfortunately, they didn't share his knowledge of chemistry. "This was a time when all these kids were blowing their hands off with match-head rockets. But I knew enough chemistry then to know that matches are dangerous, so I started working on safer fuels."

To this day, Wes remembers the common ingredients he carefully prepared for his first experiments—ingredients we will not mention here. "I'd use little twenty-two-caliber shells to start with, and pack the stuff in the shells and let it set. Then I'd put one on a scale with a fuse, fire it up, and see how much thrust I'd get out of it. Then I'd use a big thirty-thirty shell and get even more thrust out of it." Wes's methodical approach to chemistry also included safety measures. "I'd do this in the backyard because it was big, and we had a big stone fireplace. I'd use that and put the shells and scale in there, so if anything happened, it would be safe. Then I started building rockets and firing them off in the open.

"They were only little things. They'd only go up a couple hundred feet. But in my senior year of high school, the Army had a course they gave for seniors, because they were concerned about all these kids hurting themselves. They said, 'We can't stop the kids from doing this, so let's teach them how to do it right.'"

Wes quickly signed up for the Army's course, taught at Fort Belleville. There, he was introduced to aerodynamics, and the concepts of center of gravity and the center of pressure, all necessary for building rockets that fly straight, without tumbling. For that course, Wes recalls, "The final exam was to build a rocket. The teacher who ran the shop in my high school let me use all the lathes and equipment to do it. Then the Army packed the rocket to take it to Camp Lee Hill, Virginia, for our final exams. There were people from all over the country: New York, New Jersey . . ." Wes pauses in his recollection, then laughs again. "Well, not all over the country—just the East Coast. But that was 'the country' to me, you know."

When it came time for the Army technicians to fire the students' rockets, "Some worked. Some didn't. A heck of a lot of them blew up." But Wes's rocket worked . . . more or less. "The fuel wasn't packed properly and it kind of snuffed out. So it went up a thousand feet and then it stopped burning, didn't go very far."

But unlike his rocket, Wes's path was headed toward great things, and he was sure of his direction.

"I went off to college and studied chemistry, and got really focused on that. I had some vague notion of using chemistry when I got out, in some part of the space program."

In 1964, Wes earned his BS in chemistry at Brown University, then went on to Stanford University to earn his PhD in chemical physics. But though his end goal remained a job in the space program, he had no ties and no connection to it until his PhD research advisor suggested, "If you're so keen on the space program, why don't you go down to Caltech and check it out?"

Caltech—the California Institute of Technology—is one of the world's preeminent universities for engineering and scientific research and education. Among many other honors, its graduates have been awarded thirty Nobel prizes. Perhaps most notable of all its endeavors, though, is the fact that Caltech runs the Jet Propulsion Laboratory for NASA, making JPL the only NASA center that is privately managed and operated.

Wes joined JPL first as a National Research Council resident associate, and by 1969 became a permanent research scientist whose specialty was ion chemistry and planetary atmospheres—areas he was able to pursue over the course of a number of space missions, including the *Cassini* probe that arrived at Saturn in July 2004.

However, the part of Wes's distinguished career path that brought him to a pivotal position in the exploration of Mars began in 1990, when he became director of NASA's Solar System Exploration Division. It was then that all the disparate parts of what would become the history-making Pathfinder mission began to coalesce.

"The idea for the mission came out of the studies we were doing in 1990. I had just become director of the Solar System Exploration Division, and what I wanted to do was to find a way to increase the flight rate for planetary missions, because we hadn't had a launch in eleven years—from 1977 to 1989. NASA was primarily focused on the shuttle and the station at the time, and even within the context of space science, the science was focused heavily on Hubble."

One of the main reasons why no planetary exploration probes had been launched since the Project Viking Mars missions (launched 1976) and the twin *Voyager* probes (launched 1977) was because of the enormous cost associated with those undertakings. (An additional reason was the ongoing delays in launching the Galileo Jupiter probe on the only heavy-lift launch vehicle available to NASA—the Space Shuttle.)

In unofficial NASA shorthand, mission planners even talked about funding in terms of "AXAF units." AXAF stood for Advanced X-ray Astronomical Facility, one of NASA's "four great observatories": the Hubble Space Telescope, launched 1990; the Compton Gamma Ray Observatory, 1991; the AXAF, now called the Chandra X-ray Observatory, 1999; and the Spitzer Infrared Space Telescope, 2003.

In 1990, the cost of the AXAF mission stood at $1.4 billion, so that was the value of one AXAF unit. At the time, Wes says, "There was this feeling you had to have a

## MAKING A DATE ON MARS

*Choosing a Launch Window
with Porkchops*

Launching a spacecraft from Earth to Mars is like firing an arrow from a moving train at a target on another moving train—one that's traveling at a different speed and at a different elevation. Some of the variables that mission planners must account for include: the positions of Mars and Earth at the time of launch *and* the time of arrival; the speed at which Mars travels along its orbit (which varies according to Mars's distance from the Sun); the mass of the spacecraft, the speed of the spacecraft; and the desired conditions on Mars on arrival—time of day at the landing site or ability to achieve the proper orbit, line-of-sight communication with Earth, and the positions of other Mars orbiters to be used as communications relays. To add to the complexity, a change in any one of these variables causes a ripple effect that directly affects the values of almost every other variable.

To bring order to all the thousands of possible different solutions to planning a trajectory to Mars, JPL navigators use a computer program that creates a "picture" of the solutions—the fabled "porkchop" plots. These plots make it relatively easy to *see* the most optimum combination of variables. This is especially valuable because there is no one best solution. Smaller spacecraft can be sent on faster trajectories (called Type 1), because smaller spacecraft don't require as much energy. Larger spacecraft intended to orbit Mars are usually sent on longer, Type 2 trajectories. Since those spacecraft are traveling at a slower speed, they need less propellant to slow down and achieve orbit of Mars. Still, almost half the launch mass of the 2005 Mars Reconnais-

mission worth at least one AXAF unit if you were going to be a mission of any significance. There was this attitude that you had to have something big to sell well."

Faced with the desire to increase the number of missions, and knowing that the reason so few missions were flying was because of their cost, Wes saw the obvious solution: "We had to make planetary missions cheaper."

To that end, Wes created a program to come up with new, less expensive approaches to planetary exploration, so NASA could increase the flight rate even without requiring budget increases. Wes called the program Discovery.

To get things started, he assembled a team of scientists to look at what science objectives could fit within the context of this new class of less expensive missions. He added a set of experienced project managers and project engineers to see if, in fact, there *were* less expensive ways to build spacecraft and scientific instruments.

Some Discovery team members came from Caltech's own JPL in Pasadena, California, and some from Johns Hopkins University's Applied Physics Laboratory, Maryland. Wes recruited retired NASA "gurus," including Jim Martin, who had been the project manager for the Viking missions to Mars. At $4.3 billion in 2003 dollars, Viking was the most expensive planetary exploration mission ever undertaken.

At first, not surprisingly, there was no consensus. "The scientists would argue about what kind of science could be done on a cheap mission. Some said, 'Ah, you can't do anything worthwhile.' Some said, 'Sure you can!'" But once the initial shock at thinking small had passed, Wes remembers that all the participants brought a considerable measure of good faith to the process.

Eventually, "We got to the point where we thought we had some clues as to how to do this inexpensively. And the way to really understand if you can do it, is to try some examples—try to design a mission and see if that design can be built for the right price.

"What we did was take one thing that looked like it ought to be easy to do but was something we hadn't done before, so from a science perspective it would be interesting."

For that first test mission, the Discovery team chose a near-Earth asteroid rendezvous. "We'd never flown to an asteroid and actually orbited one. But the amount of energy required to get to a near-Earth asteroid—an asteroid that comes close to the Earth—is very small." Because of the enormous expense of the launch vehicle that provides that energy in the crucial first few minutes of a mission's flight, the low-energy requirement was absolutely necessary for a low-cost mission.

But the nature of scientists and engineers is to constantly push the boundaries of what might be possible, so once the Discovery team had selected their "easy" mission, Wes decided they should see if they could apply their low-cost methodology to a more challenging goal.

Wes asked himself, "What would be the most challenging thing one could do?" And he had a good answer: "Land on another planet. That would be the challenge. Haven't done that in a long time, and that's got to be pretty damn hard to do." The choice of which planet to land on was obvious. The intense heat and atmospheric pressure of Venus would make the construction of a lander too expensive, so that planet was ruled out. And of the remaining three planets that offered solid ground to land on, Mercury and Pluto were too distant to reach for the money available.

To coin a phrase, the Discovery program was going to Mars.

The Applied Physics Lab had twenty-five years of solid experience in building inexpensive orbiter missions for the Navy, so Wes assigned the near-Earth asteroid rendezvous study to them.

The Jet Propulsion Lab had recently launched the *Magellan* radar-mapping probe to Venus and, in Wes's words, that successful mission had been accomplished "on a shoestring." So Wes assigned the Mars landing study to JPL, specifically to one of the Lab's top engineers, Tony Spear. At the time, Tony was a twenty-eight-year veteran of JPL and had worked on *Magellan* for eleven of them. His positions on the mission had ranged from preproject study manager and Synthetic Aperture Imaging Radar manager, to deputy project manager, mission manager and, for Venus orbit insertion on August 10, 1990, project manager. With *Magellan* in successful operation at Venus, Tony Spear was the perfect choice to set his sights on Mars.

As Wes describes it, the initial plan for the still-unnamed mission that would become Pathfinder was dead simple: "All we want you to do is land on the planet with a Brownie, take a picture, and declare success." In this case "Brownie" refers to a simple box camera first made by Kodak all the way back in 1900.

Dull as it might seem, there was a reason for that level of simplicity. Instead of being a full-bore science mission, the Discovery Mars landing mission was to be a "technology demonstrator."

Wes explains: "What the mission was to do was to demonstrate, first and foremost, that we could do a hard mission, cheaply. And, second, if it succeeded, we would have a proven mechanism for carrying scientific packages to the surface of Mars. So for future missions, we could just build copies of the first lander and send science packages.

"The idea was to send a twenty-kilogram payload [about forty-four pounds] to the surface of Mars, so we could say the next time we do it, we can send twenty kilograms of scientific instruments."

Unfortunately, the first study back from JPL came with "a huge price tag."

Wes remembers telling the JPL team, "No, you haven't got it right. I want you to do it for *this* amount of money. I'd walk them through it, and they'd come back every time with different ways to implement the mission, but the price was always the same and way too much."

The price Wes was willing to pay, by the way, was only $150 million in 1992 dollars.

When asked how that figure was arrived at, Wes makes some comment about having consulted a dartboard, but there turns out to be logic to the calculation—it was based on the inflation-adjusted cost of the *Mariner 10* mission.

The *Mariner 10* probe was launched on November 3, 1973, and became the first—and so far, only—spacecraft to successfully reach Mercury. (The *Messenger* probe, another Discovery mission, launched in July 2004, will fly by Mercury in October 2007, again in July 2008, and finally enter into orbit of Mercury in 2009.)

sance Orbiter is the propellant it will need to slow down when it reaches Mars.

The porkchop plot shown here represents the launch opportunities for 2005. Type One trajectories are shown in the bottom porkchop, Type Two in the upper one.

The Mars Reconnaissance Orbiter is currently scheduled to launch on August 10, 2005, and arrive at Mars in early March 2006. If you trace the intersection of these two dates on the porkchop plot, you'll see they intersect almost exactly within the optimum circle of the Type 1 trajectories—in other words, bull's-eye!

—BRIAN MUIRHEAD

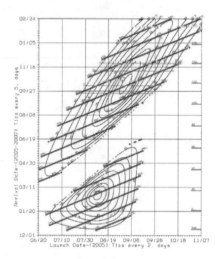

The cost of the *Mariner 10* mission in 1973 dollars was just under $100 million, so Wes added $50 million for inflation, and the $150-million price ceiling was set. Pretty scientific! (In 1992 dollars, this price did not include the cost of the launch vehicle, mission operations, or the rover.)

Another key moment in the evolution of the Pathfinder Mars mission occurred in 1990 when Wes paid a visit to JPL and received an invitation from Donna Shirley, who, at the time, was leading a technology task force looking at microrovers. Wes recalls she asked him to "come up to the seventh floor and see what they'd been doing in small rovers. Somebody had a little cart with a Macintosh computer on it that ran all around the hallway and did all kinds of stuff." The possibilities were impossible for Wes to resist, and his first thought was: "That would be neat to put on Mars."

Donna Shirley had no argument with that idea. She had known she wanted to be an engineer since the age of ten, and though her first interest had been airplanes, a few years later, science fiction gave her a new focus. She credits Sir Arthur Clarke's novel *The Sands of Mars* as having a profound effect on her choice of career, because of its depiction of a group of scientists living and working as a team on another planet.

In 1966 Donna became the first woman with an engineering degree to be employed by the Jet Propulsion Lab, and was head of the Lab's Mars exploration program from 1994 to 1998. Today, Donna has come full circle and is now director of the Science Fiction Museum and Hall of Fame in Seattle, Washington. And when Wes Huntress visited her in 1990, there was no bigger fan of putting a robotic rover on Mars than she.

Except maybe Wes himself.

After riding herd on his JPL study team under Tony Spear to control costs and keep the mission design simple, Wes drops his voice to a whisper as he admits that after he saw that Macintosh on a cart, "I succumbed to my science failing which is: try to put as much as you can on the mission."

However, as taken as he was with the idea of a rover on Mars, Wes also knew that his $150-million budget couldn't handle a lander *and* a rover. "So we started rover development, not in my division, but in another one focused on technology development, so NASA would pay for it outside the hundred and fifty."

That's when the first of many Pathfinder battles began.

The Discovery missions, were any of them ever to go forward, were technology driven, so NASA's science community had little to say about them. But the cost of developing a rover for a Discovery Mars mission was going to come out of NASA's science funding, and that *was* a plan the science community did have something to say about.

What they said was no.

As far as the scientists were concerned, science funds were being misused to develop a machine, and as Wes puts it, "They didn't want to spend science dollars doing technology. It was a fight all through this whole process with the science community, trying to keep them from undercutting the Pathfinder mission."

So, where Wes had come up with the Discovery program as a way to keep both scientists and engineers happy with an increased tempo of planetary exploration missions, he found himself being hammered from both sides.

The science community at NASA didn't want one cent of its hard-won budget allotments going to develop pie-in-the-sky technology for an all-new, never-been-done-before approach to spacecraft and mission design. "The mentality in the science community at the time," Wes explains, "said that if you're going to go somewhere, and you're going to have to spend that money, then spend as much as you can possibly get to do as much science as you possibly can. Because the next bus out of town is going to be way downstream.

"That was the kind of vicious circle we'd gotten into: don't propose a mission unless it's one AXAF unit or more, because you won't be able to get enough science. And when you do sell that, you just keep piling on as much science as you can. That's what caused growth in the mission costs and delays and more costs, and created a vicious circle. The only way to break it," Wes concludes with conviction, "was to find some way to do really inexpensive missions."

## DOING WHAT'S NEVER BEEN DONE

There's an old joke that defines an engineer as someone who can do for a dollar what any damn fool can do for two dollars.

Nowhere is that definition more applicable than in what came next in the development of the Pathfinder mission. Simply put, if the old way of doing things is too expensive, find a new way.

After several rounds of design discussions, the Discovery mission to land on Mars was beginning to acquire a shape.

The key element was that not only would part of the spacecraft land on Mars, it would deliver a rover.

Unfortunately, there was still one major problem—the price tag.

But Wes and his Discovery team at NASA Headquarters had an idea where a substantial savings might be made.

"Tony had this thing landing on rockets the old Viking way, and we said, give us a more innovative idea. Try to do something simpler because this is going to be tough to do, and you're not getting any more money!"

In 1976, the multibillion-dollar *Viking* landers had made their final descent to the Martian surface in more or less the same fashion as the *Apollo* lunar landers had set down on the Moon, just without a pilot. After shedding their protective entry shields and slowing their descent with parachutes (which the lunar landers didn't need), each *Viking* lander used three rocket engines to slow its speed to a few miles per hour, then touched gently down on three landing legs.

Without even beginning to consider the cost of developing an autonomous landing system for a new Mars lander, the fact that a rocket-powered landing would require the expense of transporting rocket engines and fuel all the way to Mars for those last tens of seconds of descent instantly added to the cost of launching the

mission. It *also* restricted the amount of payload the lander could carry—each pound of fuel required losing a pound of payload. Of NASA's first robotic *Surveyor* craft to land on the Moon, fully 60 percent of its mass was given over to its rocket landing system and fuel.

Wes admits that, at the time, he didn't know if using landing rockets really was going to be too expensive, but his opinion was that—should the Discovery Mars mission become real—the necessary technology for small rocket engines just wasn't going to be available in time.

So, Wes says, "We kept pushing Tony toward trying something those Russians did. That was pretty damn clever on *Luna 9*, you know."

*Luna 9* was the first spacecraft to successfully soft land on another body in space. In this case, the Russian lander set down on the Moon on February 3, 1966, and did so in a completely novel way.

When the space probe had descended to about fifteen miles above the lunar surface, a rubber airbag inflated around the instrument payload. The probe was oriented to fall with a sixteen-foot-long pole pointed straight down, so the moment the pole made contact with the surface, the instrument payload was ejected straight up, to fall at a much slower speed. The airbag functioned as a protective shock absorber until the egg-shaped payload had finished bouncing and come to a complete stop.

The system worked as planned, and not only was *Luna 9* the first probe to safely land on the Moon, it was also the first probe to return images from the lunar surface.

In Russia, the use of airbags to protect objects during the stresses of landing actually dates back to 1954, when very simple airbag systems were used to land undamaged aerial targets for aeronautic research on Earth. In space exploration, though, airbags were used for the 1966 landings of *Luna 9* and *Luna 13*, and then apparently forgotten.

An airbag landing system for a Discovery Mars mission offered another critical advantage in addition to not requiring large rocket engines and fuel. A lander protected by airbags could potentially set down in much rougher terrain.

Everyone at NASA recalled the first images that had come back from the *Viking 1* lander in 1976. Thirty feet away from *Viking 1* was a boulder almost the same size as the lander. The lander had no onboard intelligence to detect and avoid large rocks or sloping crater walls, so a gust of Martian wind might have been enough to spell disaster by causing the lander to attempt landing *on* the rock.

An airbag-enclosed lander, though, could have hit the boulder and bounced right off.

So Tony Spear finally agreed to investigate airbags as a landing option for the Discovery mission. But he made Wes pay a price for his agreement. Tony no longer wanted his study to be mere conceptualizing—he wanted it set firmly on the road to becoming a real mission. He asked Wes to give him "Level One requirements."

Every space mission has requirements—those conditions it has to meet in order to be judged a success or failure. They can range from the smallest detail—the

(OPPOSITE PAGE) In the days before detailed computer graphics, NASA engineers relied on traditional artists to help them visualize a mission's progress. This painting from 1974 illustrates the key events in the Entry, Descent, and Landing of a *Viking* lander as they were to occur in 1976. COURTESY NASA.

An engineering model of Russia's *Luna 9* lunar probe, the first spacecraft to successfully soft-land on another body in space *and* the first lander to use an airbag system. (The airbag is not shown.) COURTESY NASA.

range of view of an individual camera lens—to the largest goal—operate on Mars for thirty sols, aka Martian "days."

But of all the different levels of requirements that might be assigned to a mission, Level One requirements come directly from the top—NASA Headquarters. They're the ones that can't be changed and *must* be met. If at any time during the mission's design, manufacturing, and test stages it appears that a Level One requirement might not be achieved, the mission faces cancellation.

However, since the Discovery mission to Mars was "merely" a relatively inexpensive technology demonstrator, its Level One requirements turned out to be minimal. At a time when most Level One requirement documents ran well over ten pages in length, Wes recalls that the first list he gave to Tony took only half a page.

Those requirements were: demonstrate a simple low-cost system, at fixed price, for placing a science payload on the surface of Mars; demonstrate NASA's commitment to low-cost planetary exploration; and demonstrate the mobility and usefulness of a microrover on the surface of Mars.

What could be simpler?

Which brings us back to those six minutes of terror.

## THE MARTIAN DEVIL IN THE DETAILS

Let's take that simple phrase "the spacecraft will land on Mars protected by airbags" and see what happens when it's translated into reality by rocket scientists.

The date is January 3, 2004. The time is 6:54 P.M. It's cold and dark on the JPL campus in Pasadena. A warm oasis of light fills the central quad, spilling out from a clear-plastic-sided tent in which a small square of Martian terrain has been created for television crews, displaying a full-size model of a Mars Exploration Rover (MER) and its lander. The guest parking lot is crammed with media trailers and small vans from local news stations from which tall microwave antennae have been deployed. There's even a small tent set up near the visitors' center with tables full of souvenir Mars and JPL tee-shirts, mugs, and baseball caps for sale.

Throughout the Lab, hundreds of scientists, engineers, suppliers, reporters, support staff, and well-wishers are glued to video screens—if they're inside—and, if they're outside, taking care to sidestep JPL's resident herd of fearless deer.

And 105.7 million miles from the high bay in which it was assembled, almost seven months after its launch from Cape Canaveral Air Force Station in Florida, the first of two spacecraft carrying a Mars Exploration Rover is 17,000 miles above Mars, traveling at 10,000 mph.

Sometime in the next ninety-one minutes, the laws of gravity first described by Sir Isaac Newton will guarantee that the spacecraft will make contact with the Martian surface.

What remains to be seen is how many pieces it will be in when it does.

Ninety-one minutes before landing—or, as NASA calls it, L-91—the peaceful cruise phase of the first MER mission comes to an end.

Since 37 minutes, 14 seconds after its launch on June 10, 2003, the MER spacecraft has been a slowly spinning object, looking suspiciously like a 1950s flying saucer, powered by a ring of solar panels on its cruise stage. But at L-91, small rocket engines fire silently in the vacuum, and the spacecraft begins to change its position for its final descent, turning its solar panels away from the sun so that now it is powered by its onboard batteries.

The spacecraft is a little over eight and a half feet wide, and little more than five feet tall, but this gentle turn takes fourteen minutes.

The time is now L-77, and the spacecraft, completely in the grip of Martian gravity, continues to accelerate in its long fall, approaching its entry speed of 12,000 mph.

Fifty-six minutes later, at L-21, the cruise stage is jettisoned, no longer needed. For the jettison maneuver to work, a series of pyrotechnic charges have to explode powerfully enough to send a guillotine-like blade through the wires that connect the cruise stage to the entry capsule and its rover. If the charges don't ignite, or if they fail to deliver enough force to cut the cables, the remaining wires will keep the cruise stage attached, destabilizing the entry capsule, possibly causing the entire spacecraft to break apart in the Martian atmosphere.

For the complete EDL (Entry, Descent, Landing) sequence to be successful, sixty-six pyrotechnic devices will have to work flawlessly.

These first charges do what's required.

In the flight control room, the MER team now performs a critical traditional ritual to help ensure mission success: they break out their candy bars and peanuts.

We would never suggest that superstition exists anywhere within the temple to science and engineering that is JPL, but "Good Luck Peanuts" have been part of most of the Lab's critical spaceflight operations since the launch of the *Ranger 7* lunar probe on July 28, 1964. And it's not hard to see why the peanut tradition has carried on ever since.

The Ranger program's primary goal was to obtain close-up images of the Moon by having a spacecraft fly straight into it, taking pictures up to the instant of impact. Unfortunately, starting in 1961, *Rangers 1* and *2* failed to leave Earth orbit. *Ranger 3* lost contact with the Earth and missed the Moon by about 23,000 miles. *Ranger 4* hit the Moon as planned, but without ever unfolding its solar panels, which meant its cameras and other instruments never switched on. *Ranger 5* also missed the Moon, though by a smaller margin of error—this time, only 450 miles or so. And *Ranger 6* hit the Moon exactly as planned, but its cameras failed.

*Ranger 7*, though, functioned perfectly, and since it was the first *Ranger* probe launched with the benefit of Good Luck Peanuts in the control room . . . well, it doesn't take a rocket scientist to see the pattern. Or, on second thought, maybe it does.

Today, the forty-year-old tradition continues at the Lab, and for the MER landings, the peanuts are joined by candy bars. For the Mars *Pathfinder* landing of 1997, Mars Bars, naturally, were the candy of choice. But for MER, with Mars Bars no longer easily available, Milky Ways are pressed into service as an acceptable substitute. Given the pressure of the moment, no one in the control room is actually hungry for these snacks, but the tradition helps them through the tense moments remaining . . . tense moments in which they can do nothing but watch as events unfold.

Fifteen minutes later, the six minutes of terror officially begin.

At an altitude of ninety-two miles above the surface of Mars, the spacecraft has entered the outer reaches of the Martian atmosphere, first measured by the *Mariner 4* probe, almost forty years earlier.

The principles behind the technology that acts to protect the spacecraft on its descent is almost as old. The heat shield and backshell encasing the lander—together called the aeroshell—were built by Lockheed Martin Astronautics of Denver, Colorado. This company also built the aeroshell for the Pathfinder mission in 1997, using techniques developed for the aeroshells they had built for the Viking missions of 1976, derived from technology they had created for the heat shields they built for the Mercury-, Gemini-, and Apollo-crewed missions of the 1960s, based on studies first conducted during the Cold War to find ways to protect space-launched nuclear warheads from burning up before they reached their targets.

The MER spacecraft hits the atmosphere at a precisely calculated angle, heat shield first. The shield is an aluminum honeycomb structure, sandwiched between lightweight rigid sheets made of a graphite-epoxy blend used today for everything from tennis racquets to aircraft components. The outer layer of the heat shield is another honeycomb structure, this time made of a phenolic compound—essentially, a type of plastic. The honeycomb spaces in the structure are filled with what's called ablative material—that is, material designed to heat up, then burn off the heat shield, taking the heat with it. In the midst of the high-tech materials being used to protect the spacecraft, the key ingredients of the ablative material are nothing more than cork wood, tiny glass spheres, and silicone.

The heat that reaches the spacecraft is generated by the column of air the spacecraft compresses before it. It's the same phenomenon that makes a bicycle pump heat up when air is compressed inside. Technically, for everyone who remembers high school physics, the kinetic energy of the moving spacecraft is being transformed into heat energy by pushing on the atmosphere of Mars, and as the heat energy is dissipated by the loss of the ablative material, the speed of the spacecraft is reduced.

In the first four of these six minutes of terror, the aeroshell's velocity will drop from around 12,000 to 1,000 mph. By the second minute, the aeroshell will experience its highest temperatures—in excess of 2,900°F.

Now the aeroshell is a supersonic bullet, speeding through the Martian atmosphere at a steadily decreasing speed. The more it slows, the less the atmosphere is compressed, the lower the heat the heat shield is exposed to.

Finally, at about L-2, the aeroshell's speed has dropped to about 300 mph and, at an altitude of 6.3 miles, a specially designed parachute is deployed. After all the precisely scheduled events undertaken during the spacecraft's journey to Mars, the word "about" comes into use here because now the actual execution of events will be determined by the spacecraft's onboard computer. It is busily measuring the aeroshell's deceleration, and estimating its speed to choose the right moment—about Mach 2 (twice the speed of sound on Mars)—to fire the mortar charge that will release the parachute. If it's deployed too early, the aeroshell's speed could rip the parachute's fabric to shreds. If it's deployed too late, there won't be enough time for it to slow the spacecraft to a safe landing speed. In the case of this landing, the computer deploys the parachute a few seconds later and about a mile lower than expected, but still within the margin of safety.

About twenty seconds later, with the parachute fully open and continuing to slow the aeroshell's descent, more pyrotechnic charges fire, and the scorched heat shield drops away from the backshell, leaving the lander exposed.

Ten seconds later, now about one minute and twenty seconds from landing, still more pyrotechnic charges fire to let the lander drop sixty-five feet down from the backshell on a ribbonlike cord.

At L-35 seconds, the lander is 1.5 miles above the Martian surface, and a small radar unit switches on to precisely determine the altitude and measure the rate of descent, giving the computer the critical information it needs to help determine the timing of the final steps in the EDL sequence. (Like everything else in EDL, no radar signal, no safe landing.)

At L-30, L-26, and L-22 seconds, the spacecraft's DIMES system comes to life.

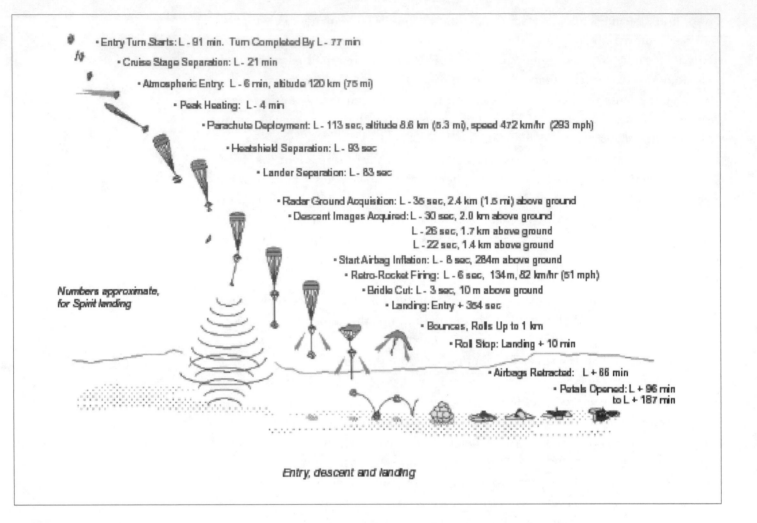

- Entry Turn Starts: L - 91 min.  Turn Completed By L - 77 min
- Cruise Stage Separation: L - 21 min
- Atmospheric Entry:  L - 6 min, altitude 120 km (75 mi)
- Peak Heating:  L - 4 min
- Parachute Deployment: L - 113 sec, altitude 8.6 km (5.3 mi), speed 472 km/hr  (293 mph)
- Heatshield Separation: L - 93 sec
- Lander Separation: L - 83 sec
- Radar Ground Acquisition: L - 35 sec, 2.4 km (1.5 mi) above ground
- Descent Images Acquired: L - 30 sec, 2.0 km above ground
    L - 26 sec, 1.7 km above ground
    L - 22 sec, 1.4 km above ground
- Start Airbag Inflation: L - 8 sec, 284m above ground
- Retro-Rocket Firing: L - 6 sec,  134m, 82 km/hr (51 mph)
- Bridle Cut: L - 3 sec, 10 m above ground
- Landing: Entry + 354 sec
- Bounces, Rolls Up to 1 km
- Roll Stop: Landing + 10 min
- Airbags Retracted:  L + 66 min
- Petals Opened: L + 96 min
    to L + 187 min

Numbers approximate,
for Spirit landing

Entry, descent and landing

The sequence of planned critical events that occurred during the Entry, Descent, and Landing of the Mars Exploration Rovers *Spirit* and *Opportunity*—otherwise known as the "simple" way to land on Mars! COURTESY NASA/JPL.

The acronym stands for Descent Image Motion Estimation System, and its key component is a small black-and-white camera that takes three images of the ground directly beneath the lander at four-second intervals. The last image is captured with the lander less than a mile above the surface, while falling at over 150 mph.

If everything goes well, when those DIMES images are transmitted to JPL sometime after landing, the features in them will help the MER team locate the lander's precise position on Mars. For now, though, the onboard computer analyzes the differences in the images to determine the lander's horizontal speed, while the radar continues to measure the vertical speed.

At L-8 seconds, the lander is 932 feet above the surface, and this is when the airbags—remember them?—inflate.

In a process similar to the system used to instantly expand passenger-safety airbags in cars, three gas-generator cartridges, each looking remarkably like a stainless steel kazoo, fire into a chamber containing three highly reactive chemicals.

That chemical reaction creates a cloud of gas that expands at 200 miles per hour to fill the four interconnected airbags surrounding the tetrahedron-shaped lander in a fraction of a second. (A tetrahedron is a four-sided pyramid shape.) While a typical car tire might be inflated to a pressure of about 30 pounds per square inch, the

airbags surrounding the lander only require 1 pound of pressure, but still are rock hard to the touch.

Two seconds later, at L-6 and an altitude of 342 feet, three deceleration rockets on the backshell, still 65 feet above the dangling, airbag-encased lander, fire to reduce the lander's descent to as close to 0 mph as possible. At the same time, because of the computer's assessment of a dangerously high horizontal velocity (based on the DIMES images), a sideways-mounted rocket on the backshell also fires, part of what engineers call TIRS, for Transverse Impulse Rocket System. The TIRS rocket imparts just enough motion to the backshell that while the RAD rockets fire, the horizontal motion of the lander at the end of its bridle is brought close to zero as well.

And finally, after a near seven-month journey of more than 300 million miles, the MER-A lander has reached L-3—three seconds from Mars touchdown—and a virtual dead stop.

As originally planned, the MER flight controllers had expected the lander to be

The airbag configuration for the Mars *Pathfinder* lander (shown here) was adapted with more layers and stronger fabric for the heavier lander of the Mars Exploration Rovers. What appear to be six spherical airbags arranged in a triangle are actually six lobes of a single airbag—one airbag to each of the lander's four sides. The airbags for all three Mars lander missions were made by the ILC Dover Company of Frederica, Delaware, which has been manufacturing spacesuits for NASA since Project Apollo. COURTESY NASA/JPL.

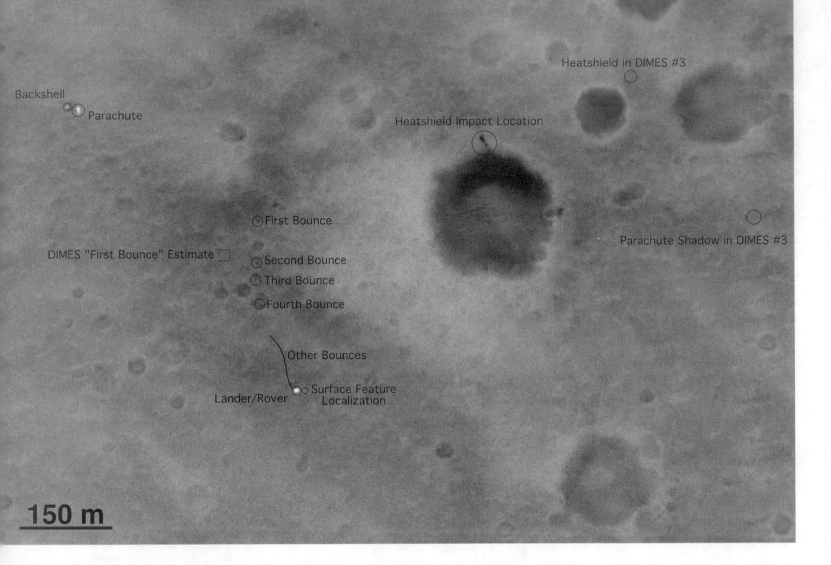

Backshell

Parachute

Heatshield in DIMES #3

Heatshield Impact Location

First Bounce

DIMES "First Bounce" Estimate

Second Bounce

Third Bounce

Fourth Bounce

Parachute Shadow in DIMES #3

Other Bounces

Lander/Rover

Surface Feature
Localization

**150 m**

The record of *Spirit*'s landing on Mars is preserved in this image returned about two weeks
later by the Mars Global Surveyor. The two dots
in the upper left are the spacecraft's backshell
and parachute, released when *Spirit*'s bridle
was cut, allowing the lander to bounce to a rest
while safely encased in airbags. To the far right
of the image, a dark streak above a large crater
shows where the heat shield hit, after having
been jettisoned several miles above the surface. The trail of bounce marks made by the
airbags can be seen in the middle of the image.
To the left of the second bounce mark is a
square showing the location where engineers
had initially estimated the lander first hit the
Martian surface. That estimation was based on
images from the Descent Image Motion Estimation System (DIMES). The white dot near the
bottom of the image is the lander itself. COUR-
TESY NASA/JPL/MALIN SPACE SCIENCE SYSTEMS.

39 feet, 5 inches above the surface. But subsequent analysis of measurements taken by
its computer will show that the lander is at 27 feet, 11 inches, and it is now that the last
EDL pyrotechnic charge fires to cut the lander's bridle, so it once again falls free as the
continuing blast from the descent rockets carries the backshell and the parachute far
away to avoid draping the lander and rover with a potentially inescapable net.

Six minutes ago, encased in its aeroshell, the lander was traveling at 12,000
mph. Now it hits Mars at less than 20 mph. Protected by the inflated airbags, it behaves like a huge beach ball and *bounces.*

The first bounce takes the lander back up to a height of 27 feet, 7 inches—
almost as good as a SuperBall!

And then the lander keeps bouncing—twenty-seven more times over the next
fifty-seven seconds.

When the lander's motion sensors tell the computer the lander has come to a
complete stop, yet *another* EDL sequence begins.

First, the lander's winches come to life, reeling in cords inside each airbag. Four
of these cords are attached to flaps that cover vent holes on the inside of each airbag.

When each flap is pulled back, the gas escapes through the holes and the airbags deflate.

For the next hour at least, various motors within the lander will continue to operate. Some will reel in the complex network of cords to retract the airbags until they're safely under the lander. Others will open up the three "petals" that have been folded up around the Mars Exploration Rover securely attached to the lander's base. (The trick is to package *everything* inside!)

The petals open in such a way that no matter on which of its four sides the lander has come to rest, by the time all three are fully opened, the base will be down, leaving the rover upright and ready to roll. This is the beauty of the four-sided pyramid lander.

Then, over the next seven to ten days, yet another complex series of mechanical movements and the firing of even more pyrotechnic charges will allow the rover to "stand up" and detach itself from the lander.

Perhaps there is no greater insight into the world of engineering than realizing that when it comes to landing on another planet, this is about as simple, low-cost, and reliable as it gets.


## BACK TO THE FUTURE

Three weeks later on January 24, 2004, the fantastic success of the landing of MER-A, carrying the rover called *Spirit*, was matched by the equally successful landing of MER-B, carrying *Opportunity*, on the opposite side of Mars.

Once again, a spacecraft jettisoned its cruise stage, hit the Martian atmosphere at 12,000 mph, successfully ignited 66 separate pyrotechnic devices, cut cables, inflated airbags, and bounced to a safe landing—this time scoring an interplanetary hole-in-one by coming to rest *inside* a crater whose exposed rim would soon provide our most profound discoveries yet about Mars.

To say the engineers who had designed, built, and tested the MER spacecraft were elated would be an immense understatement. To say the scientists who could now use *Spirit* and *Opportunity* to truly explore where no one had explored before were thrilled, would equally misrepresent the level of excitement, relief, and anticipation that filled JPL and spilled out to the world.

Consider the context of this success.

Eleven months earlier, NASA—and America—had suffered a horrendous human tragedy when the Space Shuttle *Columbia* disintegrated on reentry, losing all seven astronauts onboard and the nation's sense of direction for human spaceflight.

Just over five years earlier, NASA had endured a professional and public-relations disaster when two Mars probes—the Mars Climate Orbiter and the Mars Polar Lander—both were lost on arrival at Mars because of errors that could—and should—have been discovered and corrected well before the failures occurred.

In the last weeks of December 2003, as the MER spacecraft made their final approaches to Mars, both Dr. Ed Weiler, NASA's associate administrator for Space Science, and Dr. Charles Elachi, director of JPL, joked they'd be taking up new careers as sheepherders in New Zealand if both spacecraft failed.

But the sober truth both knew was that failure would mean NASA's current program for the ongoing exploration of Mars would effectively be canceled, and it could be years before confidence could be rebuilt and Mars exploration could begin again. The future of JPL, if it had a future at all, some believed, hung in the balance. This was a "bet your laboratory" moment for the JPL director.

No wonder then that failure was on everyone's mind.

NASA public affairs officials even began to prepare the media for the worst possible scenario, but the media had already anticipated them. In the week before the arrival of *Spirit*, newspapers and television reporters ran story after story calling Mars "the Death Planet."

After all, some of the stories explained in prose more suited to 1930s pulp science fiction than twenty-first-century science reporting, of the forty-two missions sent to Mars since the launch of Russia's *Marsnik 1* in 1960, as of December 2003 only fourteen could be considered successful, four partially successful, and the other twenty-four outright failures. That's a success rate of only 33 percent! The Internet buzzed with whispered tales of Martian conspiracies and something called "the Martian Grinch"—a strange, uncharted phenomenon near Mars that was responsible for the inexplicable destruction of spacecraft after spacecraft, almost (cue the creepy theremin music) as if someone—or something—didn't want us to go there . . .

But as the old saying goes: There are three kinds of lies: lies, damned lies, and statistics.

So let's take another look at Mars's reputation as a death planet.

First of all, we'll drop the Russian probes from consideration. To be fair, Russia led the space race for many years and achieved many notable firsts: first man in space, first woman in space, first probe to land safely on the Moon, and first space station, among others. Today, the Russians are an invaluable part of the effort to maintain the International Space Station, their rocket boosters have an outstanding performance record, and their robust *Soyuz* spacecraft serve as the space station's emergency lifeboats. But in the '60s and '70s their approach to engineering under the communist system didn't lend itself to success.

So if we look at the scorecard for only NASA's missions to Mars, suddenly we're in a whole new ballgame.

From NASA's first attempt to send a probe to Mars—*Mariner 3* in 1964—till the arrival of the successful *Mars Odyssey* orbiter in 2001, NASA undertook seventeen missions to Mars, of which twelve were completely successful. That's a success rate of 70 percent!

If we exclude NASA's early flyby attempts when disaster often struck in the first few minutes after launch, and examine only attempts for the technically more challenging Mars lander missions, then that success rate climbs to seventy-five percent: out of four attempts—two *Viking* landers, Pathfinder, and the Mars Polar Lander—NASA succeeded three times. (Russia has attempted to land on Mars six

When *Spirit* suffered a temporary computer glitch early in its mission, this image—supposedly the last one sent back before the glitch began—quickly spread through the Internet as a possible explanation for the rover's unexpected (and quickly corrected) troubles. Far from being a Martian Grinch, though, the Warner Bros. cartoon character Marvin the Martian served as the unofficial mascot to the Mars Pathfinder mission. For the MER missions, Marvin and his archrival Duck Dodgers became "official" mascots of the Delta launch teams. Look for their mission patches in the color section of this book. ARTIST UNKNOWN/MARVIN THE MARTIN™ & © LOONEY TUNES & WARNER BROS.

times. The only lander that managed to accomplish that goal—*Mars 3*, December 2, 1971—remained functional for a total of twenty seconds and returned no useful information. It was, however, the first successful soft landing on Mars.)

Then, as of January 24, 2004, when the first images came back from *Opportunity*, the second rover to land on Mars that month, NASA's landing success rate skyrocketed to 83 percent—five for six!

So no more talk about death planets or Martian Grinches, please.

There will always be a significant element of risk in space exploration, if only because so much of what NASA sets out to do is something that has never been done before.

But in the case of landing the *Spirit* and *Opportunity* rovers on Mars, the MER teams responsible for that double-barreled success did have the advantage of following in the footsteps of the Pathfinder team, who had literally blazed the trail through the Martian atmosphere six and half years earlier.

In fact, several key members of the Pathfinder team were reprising their roles on MER—so they weren't just going to Mars, they were going *back*.

## AELITA: QUEEN OF MARS

*Setting the Stage*

As a science-fiction film, the 1923 Russian silent movie *Aelita: Queen of Mars,* had a profound influence on science-fiction illustrators and many American films to follow, especially the Flash Gordon serials in which the sets and costumes of Planet Mongo echo those in this film. However, as a film about Mars, *Aelita* chiefly serves to remind us today how mysterious that planet was in the 1920s. With only blurry photographs to go by, and the popular belief that some form of vegetation, and perhaps even people might exist there, to the creators of *Aelita,* Mars became a place where anything could happen—and did—because no one would be able to prove otherwise.

Though it is one of the first major film productions to be set on Mars (with a cast of 3,000!), *Aelita: Queen of Mars* is more a political fantasy than an attempt to portray a possible version of the planet. It was based on a 1923 Russian novel by Count Alexei Tolstoi (a distant relation of the more renowned Russian novelist Leo Tolstoy), and scientific accuracy was the least of the writer's concerns. In the novel, the journey from Earth to Mars in an egg-shaped, rocket-propelled craft takes about ten hours, compared to the average seven- to twelve-month journey required by today's spacecraft.

In the film, the hero travels through space in a metal spaceship shaped like a hot-air balloon, but the craft is seen only briefly as it takes off from the outskirts of Moscow, and again when it crash-lands on Mars.

The landscape of Mars itself takes up even less screen time, merely a brief glimpse of a miniature set depicting an agricultural area bordered by rocky hills. The real action of the film takes place inside spectacular Martian buildings—including the Tower of Radiant Energy—and in underground slave-labor installations. Indeed, Mars has never been more aptly named the Red Planet than in this film, featuring one of the Earthlings leading the Martian workers in a revolt which has them forging hammers and sickles to proclaim a Martian Union of Soviet Socialist Republics!

# CHAPTER TWO

# "You Don't Want to Get Too
# Close to These Guys"

## Assembling the Team

*On the afternoon of October 19, 1899, I climbed a tall cherry tree at the back of
the barn . . . As I looked towards the fields to the east, I imagined how wonderful
it would be to make some device which had even the possibility of ascending to
Mars . . . I was a different boy when I descended the ladder. Life now had a purpose for me.*

—ROBERT H. GODDARD, "FATHER OF MODERN ROCKETRY"

## THE MISSING MAN

Going to Mars is not just about engineering.

The date is July 4, 1997. The place, once again, is JPL, where it is 8:30 in the
morning. But about 120 million miles away, for a little spacecraft falling toward
Mars—it's L-90.

The Entry, Descent, and Landing sequence that will be used in the future by the
MER spacecraft of 2004 is about to begin for the very first time.

And something's gone wrong.

Keep in mind that nothing quite like this had ever been done before.

NASA's last lander mission to Mars had taken place twenty-one years earlier,
when the agency's approach to planetary exploration was significantly different.
The twin *Viking* spacecraft of 1976 were fantastically complex and staggeringly expensive machines. In 2004 dollars, the twin MER spacecraft had cost $615 million
to develop. Adjusted for inflation, the twin *Viking* spacecraft had cost the equivalent

of $3.9 *billion* to develop. Today, spacecraft of that cost and complexity are sometimes, tongue in cheek, called *Battlestar Galacticas*.

True, each *Viking* probe consisted of an orbiter and a lander, but the mission design itself contributed to the astonishing cost.

In 1976, both spacecraft had launched from Earth, traveled to Mars, and then had gone into orbit around the Red Planet for an almost leisurely few weeks during which potential landing sites were imaged in detail and final decisions had been made back on Earth.

The cost and complexity of orbiting a lander before committing it to a landing, to say nothing of the energy cost in transporting enough rocket fuel to Mars to allow the landers to slow to orbital speeds, was *exactly* what Wes Huntress was striving to eliminate with his new Discovery program. He was determined to create simpler, less expensive missions.

The *Pathfinder* spacecraft falling toward Mars in 1997 *was* that completely new approach.

*Pathfinder* wasn't going to orbit Mars—no time for looking for just the right landing site. It was going to hit the Martian atmosphere at a full-bore 17,000 mph, shedding its speed by means of an ablative heat shield, then a parachute, then a few seconds' worth of rocket fire from its backshell retros, and then absorb all the remaining energy of its plunge with airbags. Like *Viking*, yet oh so different.

Where the MER flight control teams in 2004 would have to face six minutes of terror with the knowledge that at least the concept of what they were attempting had been proven successful, the Pathfinder flight team was facing the five minutes of terror (their spacecraft was traveling faster) with *no* preceding example to look back on.

The first challenge for *Pathfinder* Entry, Descent, and Landing was the accuracy its navigators were attempting. Though Mars was "only" 119 million miles from Earth on July 4, 1997, the little spacecraft had traveled 309 million miles because it had been launched toward a moving target. Now it had to hit that moving target with an angle of atmospheric entry accurate to within only one degree. Too shallow an angle, and the spacecraft would skip back into space like a rock skipping across a pond, to be lost forever into orbit around the Sun (possibly to encounter Mars again in the distant future). Too steep an angle, and *Pathfinder* would become a flaming fireball hurtling to the Martian surface, making a crater instead of exploring one.

One member of the Pathfinder navigation team, Robin Vaughan, did the math to make the comparison between *Pathfinder*'s trajectory to Mars and a golf game. Robin set Cape Canaveral—*Pathfinder*'s launch site—as the "tee," and her "green" became that portion of the Martian atmosphere the spacecraft could enter at the proper angle and speed without being destroyed. That safe target area had a diameter of twenty-six miles. Then, for the spacecraft to have a chance of landing in the designated target site on the surface, not only would it have to reach the green, it would have to precisely pass through the "hole." In this case, that was a section of the green that was only fifteen miles in diameter.

Bringing those calculations back to Earth, where a golf hole is four and a quarter inches wide, Robin worked out that if the *Pathfinder* trajectory to Mars were a golf shot, the tee would be in Pasadena, California, the four-and-a-quarter-inch hole would be just outside of Houston, Texas, and the surrounding green would be all of seven inches wide.

Tiger Woods, eat your heart out.

In addition to navigational accuracy, one of the reasons why sending probes to Mars is so technically challenging is because, at that distance, commands sent from Earth to a spacecraft at Mars can take anywhere between three to twenty-two and a half minutes to reach the Red Planet, and that's traveling at the speed of light—186,000 miles per second. Then, it takes just as long for the spacecraft to send back its acknowledgment.

With that kind of communications delay, there is no method by which people on Earth can directly and immediately control a spacecraft on Mars, whether in orbit or on the Martian surface. And that means the spacecraft has to have enough onboard computing smarts—in hardware and software—to control its *own* behavior in response to rapidly changing conditions; for instance, the conditions *Pathfinder* faced as it decided when to open its supersonic parachute deep in the Martian atmosphere.

Three weeks before landing, the Pathfinder flight controllers had transmitted to the spacecraft a new load of software, a new lesson in how to enter, descend, and land on Mars. Unfortunately, despite rigorous simulations on Earth, there was no absolute way for *Pathfinder*'s navigators to determine whether the spacecraft had truly learned its new lessons before those lessons would be put to the test.

For *Pathfinder* at L-90 minutes before landing, its Entry, Descent, and Landing final exam began as the spacecraft took its first actions solely under its own control. Up to this point, the operators on the ground had a chance to step in to correct any mistake the spacecraft might make. But now they had no choice but to trust the onboard software.

*Pathfinder*'s first autonomous action was to vent Freon gas to space. In the spacecraft, Freon served the same purpose as in an ordinary refrigerator on Earth—it circulated through a network of cooling tubes and radiators to keep the electronics deep inside the lander from getting too hot in the insulated aeroshell. But now that the cruise stage needed to be jettisoned, the Freon had to be vented in a controlled way or it could disturb the spacecraft's attitude at entry and put the entire EDL sequence at risk.

In their small flight control room at JPL, with Sun workstation computer screens the only window on what was happening on their spacecraft, Brian Muirhead and the Pathfinder team watched anxiously to see if this venting disturbed the spacecraft's precise attitude. It didn't. The first step had been successful. Only another fifty-three to go.

The spacecraft then started to execute a number of other programmed instructions to make itself safe for its fiery entry. Just like pilots going through a checklist, confirming airspeed and setting flaps before landing, the flight team watched their screens as the signals sent from the spacecraft let them know that *Pathfinder* was doing everything it was supposed to . . . until it reached a point in the EDL sequence where two valves in the propulsion system were supposed to close.

According to the signals—what NASA calls telemetry—those valves remained open.

Questions began firing around the control room. "Weren't those valves supposed to close? Can anyone verify that the latch valves are closed?"

On the one hand, whether the valves were open or closed didn't really matter all that much. What was alarming the team was that the new software hadn't done what it was supposed to do. And that raised an unsettling question . . .

It was up to Richard Cook, *Pathfinder*'s mission operations manager, to ask the question aloud.

He turned from his computer screen to Rob Manning, *Pathfinder*'s flight system chief engineer, and asked, "Are we *sure* that the new load of software is onboard and running?"

If it wasn't, Richard knew, if the software upgrade transmitted to the spacecraft three weeks earlier had somehow failed to load, then it was possible that *Pathfinder* was going to hit the Martian atmosphere with the original program, without the fixes and upgrades needed for a safe entry and landing.

Now Richard and Rob and the flight software leads, Glenn Reeves and Steve Stolper, studied the screen that displayed the latest spacecraft telemetry—tiny columns of words and numbers that tell the experienced reader everything he or she needed to know about the health and activities on *Pathfinder*.

They looked near the bottom of the screen to check the software version number. The date the new code was completed and compiled, R061097, was correct. So what was wrong? The pitch of people's voices began to rise as they called out, "Is there anything else wrong? Could the version number somehow be wrong? Was there some other way to be sure that the new code was running?"

But all the changes the team had made in the new programming affected only spacecraft activities that were still to happen. There would be nothing on the screens that could tell anyone that the software was actually doing any part of its new lesson.

The team faced an agonizing decision. With EDL just over an hour away, there was *technically* only one option: to send a reset command and cause the spacecraft computer to shut down and restart—exactly like you'd do when your home computer gets flaky or freezes up. But if *Pathfinder*'s computer didn't come back up, it would be the end of the mission, and the mission managers would have to explain to the failure review board why they did such a stupid thing . . .

Then Richard Cook had a brainstorm. "Wasn't the fill packet changed in this new load of software?"

The excited answer was, "Yes!"—given by software engineer Steve Stolper, who immediately ran out of the room.

Steve had rushed off to look at the content of the most recent fill packets.

"Fill packet" is a term used by JPL software engineers, and a good way to picture them is as the wads of cotton batting that get jammed into pill bottles to keep the pills from rattling by tightly filling the bottle to the top.

Telemetry sent from the spacecraft to Earth is sent in fixed-size chunks called "frames." Now, it often happens that the data, for example a picture or part of a picture, don't fill the frame just as fifty aspirin might not fill up a pill bottle. So the flight computer is programmed to add the right amount of "fill" to the frame before it is sent to the ground.

The usual content of a fill packet offers insight into classic software engineer "geek" humor. For *Pathfinder*, typical fill packets contained names of the team members plus important messages like team members' names, famous quotes, biblical verse, and "Elvis lives."

But the content of the particular packet that Steve Stolper ran out to look for was special, a certain name: Jordan Kaplan.

Jordan Kaplan had been a popular member of the Pathfinder team—a young engineer with degrees from Tufts and MIT who had helped build the spacecraft's lander. Like most engineers at JPL, space exploration was only one of his many interests.

He loved aviation and flying, and had his own antique 1946 Ercoupe single-engine plane.

He was an accomplished violinist and performed classical music with the Santa Monica Symphony Orchestra.

And on April 18, 1997—two and a half months before *Pathfinder*, the lander he had worked so hard on, was to reach Mars—Jordan had taken off in his beloved plane to fly to a concert where he would perform, and the plane's engine caught fire.

Jordan's passenger died in the crash. He was severely burned.

The JPL community banded together to support Jordan and his family, hoping for his recovery, but his lungs had been too badly damaged. Seven days later, on April 25, Jordan Kaplan died.

It was a devastating loss for the *Pathfinder* team, and Brian Muirhead asked Glenn Reeves, flight software lead engineer, to find a way to put Jordan's name on the spacecraft as a tribute to their lost friend.

Glenn's solution was to put Jordan Kaplan's name into the fill packet, so he would be there when the spacecraft communicated with home.

While the rest of the operations team racked their brains over what could have caused the propulsion valves to not close, Steve Stolper flipped through pages of raw telemetry stored in the project database, looking for that one name.

Brian was standing next to Richard Cook when Steve came running back in. "It's there, it's there! Jordan's name is there!" Richard turned to Brian. "Thanks for putting Jordan's name on the spacecraft." "Thank you," Brian said, "for knowing our spacecraft so well."

In the end, no one ever determined if the valves were open or closed, but their condition wasn't the reason for the team's concern. Science is all about prediction, and engineering is all about designing and building devices and computer programs and processes that behave in predictable ways.

The fact that the valves' condition did not fit into the predictions that had been

On landing day, beside a model of the *Pathfinder* lander and *Sojourner* rover, Brian Muirhead, NASA Chief Administrator Dan Goldin, and JPL Director Dr. Ed Stone receive a congratulatory call from Vice President Al Gore.

made for *Pathfinder*'s EDL suggested that the complex system had not been thoroughly understood, which in turn suggested that other unpredictable events could occur. And at a speed in excess of 17,000 mph, a single unpredictable event could mean disaster.

But the reassuring presence of Jordan Kaplan's name told the Pathfinder team that they had done everything they were supposed to have done—the new software was at least onboard their spacecraft and there was no need to do anything drastic.

This whole crisis had played itself out in what seemed like an eternity, though it had been only twenty minutes. There was still an hour to go.

But Jordan Kaplan's presence in the memories of the Pathfinder team, and in the faint telemetry signals coming from Mars, made what could have been an hour of total panic just a little easier to bear.

*Pathfinder*'s mission was to put a robot on Mars, but it was still a very *human* endeavor.

## CHAOS THEORY

The Pathfinder mission to Mars began at NASA Headquarters with Wesley Huntress Jr. and his Discovery program to develop a new and less expensive approach to planetary exploration.

At JPL, Tony Spear and the Pathfinder team took on the task of developing the mission to use airbags to land a small rover on Mars and take a picture, just to prove that the next time, fifty-one pounds of scientific instruments could be sent for the bargain-basement price of only $150 million. (In 1992 dollars, not including the rocket and operations, of course.)

And seven years from the time *Pathfinder* was a glint in Wes's eyes, that mission was accomplished, becoming in the public eye NASA's biggest planetary space exploration event since the first human landing on the Moon.

There are many reasons for that success—including great engineering and, everyone must admit in hindsight, a bit of luck. (It is sometimes said that JPL stands for "Just Plain Lucky." But with forty-two events having to happen at precisely the right time during EDL, there's a lot more to it than luck.) But the most important factor in the success of *Pathfinder* was the unusual group of people who came together in the right place at the right time to make history, and even a little bit of magic.

Why unusual?

Because the whole project was unusual—combining a bit of the Apollo-era "anything's possible" attitude mixed with some completely new ways of doing business. And because, for the Discovery program to work, it was inherent to the process that the team responsible for any Discovery mission would have to take RISKS.

Think about that for a moment.

Engineers are driven by the demands of the business to be very accurate and strive for perfection in every way possible.

Under NASA, deep space exploration is driven by scientists and implemented by engineers, and in 1990 those scientists and engineers were focused on billion-dollar-plus missions because only that kind of money could produce fully redundant, very high reliability spacecraft designs, guaranteed (as much as anything can be guaranteed in the aerospace business) to take scientific instruments where they needed to go.

The downside to that approach is that a single billion-dollar mission would suck up all available funding for years, leaving NASA in the position of putting all its scientific eggs in one basket. That meant that a single failure in a billion-dollar mission could effectively wipe out a significant fraction of NASA's scientific efforts for those years, not to mention igniting a proverbial firestorm of media and public criticism.

But if the Discovery program worked, then Wes Huntress could see $1 billion supporting as many as *five* missions—five teams of engineers, five teams of scien-

## FOLLOW THE MONEY

*How Planetary Space Science Missions Are Funded Within NASA*

Unlike Athena, daughter of Zeus, space science missions do not spring fully grown from the heads of program managers (or scientists). There is a long and often tortured process from concept to formulation to development and finally into operations. Each step has different stakeholders and funding constraints, and they can change as quickly as the weather on Mars.

The process may begin with a specific call for proposals associated with an existing program like Discovery, from which the Stardust, Genesis, Deep Impact, and Messenger missions came. (Those missions, respectively, are a comet coma dust sample-return, a solar wind sample-return, a cometary impact and observation, and Mercury orbiter.) Or it may be a directed mission from NASA Headquarters, such as Mars *Pathfinder,* the Mars Exploration Rovers, the NEAR asteroid rendezvous mission (built by the Applied Physics Lab at Johns Hopkins University), the *Galileo* spacecraft and probe to Jupiter, and the *Cassini* spacecraft and *Huygens* probe (provided by the European Space Agency) to Saturn. The trend today is toward more competitively selected missions that are aligned with the NASA's ten-year scientific strategy called the Decadal Survey.

The competitive process starts when NASA issues a call for proposals. This is paid for by the internal funds from the institution and industry partner making the proposal and by lots of unpaid overtime by the proposal team who develop a technical, design, and programmatic plan. The resulting overall plan covers a project development cycle that is generally talked about in five phases, from A through E.

The broad constraints for a project are often known prior to much technical work being done. In the case of existing programs such as Discovery or New Frontiers, there is a development schedule of about three years and a defined cost cap. The cap for Discovery is currently about $350 million, and for New Frontiers, about $700 million. The proposal team's job is to fit a scientifically valuable/exciting mission into the cap with a "sellable" implementation approach including "adequate" budget reserves and "manageable risk."

In the case of directed missions from Headquarters, the cost cap may be "hard" or "soft." The key constraint is the NASA budget, which is planned five years ahead with funding allocations for existing programs and start-up funding wedges for new ones. As there is currently no reserve held at NASA for space science missions, the money needed to make up any cost overruns comes out of the hides of other ongoing or future projects, sometimes resulting in scaling back the scope of a mission or delaying its launch date. Even in space, it's truly a dog-eat-dog world.

The first gate for a project to pass through into Phase A is either winning the first round of a competition, completing a particular review, and/or getting a letter from NASA Headquarters providing guidelines for the initial phase of work. In any case, the necessary ingredients to start Phase A are high-level mission and scientific requirements and some money. Funding at this preliminary level could range from hundreds of thousands to tens of millions of dollars.

In Phase A, the project develops its initial requirements, conceptual design, and preliminary cost estimate. This work culminates in a formal review in which NASA has the option, based on technical and programmatic considerations, to continue funding, redefine the mission, or kill it. If you survive this review, you're sent a letter authorizing you to proceed to Phase B. At this point, the required staffing and available funding take a significant step up.

In Phase B, the mission requirements are solidified, the design is further detailed, implementation plans are formalized, and a detailed formal cost estimate developed. In some cases, advanced procurements may be allowed for special equipment, such as specialized components, electronics, and science instruments that require a long lead time to design and build.

Phase B culminates in the Preliminary Design Review, called the PDR. If the project passes its PDR in good technical position and with a sound budget and risk-management plan, the NASA Associate Administrator for Space Science may bestow upon the team the coveted status of an approved flight project, which translates to: "Welcome to the real world—Phases C and D." Upon receipt of the official letter of confirmation from Headquarters, we pat each other on the back and say: "Congratulations, they called our bluff, now we have to go do it!"

tists, turning NASA's expertise loose on five different targets in space. And with five missions flying, if there were a single failure that doomed a probe, why then, there'd still be four missions alive and delivering results.

In other words, by accepting a reasonable amount of risk for considerably less in the way of taxpayer dollars, NASA could conceivably increase its science return in the face of a failure.

In Phase C, which is the detailed design and fabrication phase, the real money starts to flow. Two thirds to three quarters of the development budget is planned and spent in this period. This is also where most of the "I forgot"s and "Oh, @#*!"s show up, including the sticker shock for contracted items. Midway through this phase, a formal review of the detailed design, called the Critical Design Review, or CDR, is held to approve commitment to fabrication. At this point, the project's spending is peaking, and almost without exception, in spite of all the best-laid plans, it's generally beyond its available funding. If spending is higher than planned, and/or reserves have been depleted beyond acceptable levels, then your project could be up for a "Cancellation Review."

In spite of the fact that a project may already have spent a significant portion of its funding, the threat of cancellation for cost overruns always looms. The threat is real, but has only been applied rarely, which is a good indication of NASA's commitment to the responsible oversight of its funding and of its appreciation of the difficulties of predicting how a high-tech development will proceed more than three years in advance. Assuming you have convinced the NASA program managers of your legitimate need for additional money and aren't going to be canceled, additional funding is somehow found, allowing your project to proceed.

Phase D is known as the Integration, Test, and Launch phase. This begins with the delivery of flight subsystems, including electronics, structure, cabling, and software to the spacecraft assembly and test team. This period can be as long as half the total development period—[eighteen out of thirty-six months for Mars *Pathfinder,* thirteen out of twenty-seven months for the Mars Exploration Rovers]. Here, all previous sins become manifest—at least, we fervently hope they do. This process ends with a Launch Readiness Review where the completeness and performance of the test program is evaluated. If all goes well, for a JPL mission the JPL director then makes a recommendation to launch to the NASA Associate Administrator, and the final decision to "GO" becomes NASA's call.

The last phase, Phase E, is Operations. Technically this phase starts the instant the spacecraft leaves the launchpad, though we don't actually start communications and control until the spacecraft is released from the launch vehicle's final stage (and we don't start the new funding until about thirty days after the launch). This phase has a predefined period for meeting the mission objectives, measured in months or years, including many critical events, such as entering orbit or landing, and returning, synthesizing, and publishing the new scientific information that was the reason for undertaking the mission in the first place.

It is the case for most successful missions that they are extended beyond their planned mission life to allow further science investigations. *Voyager 1,* launched in 1977, started being extended after it passed Saturn in 1980, and is still being monitored today, almost thirty years later. With each extension, additional funds must be provided to cover the cost of communications and the ongoing work of engineers and scientists assigned to operate the mission and process the incoming data.

For surface missions operating on Mars time, extended missions are a mixed blessing. More science means more wear and tear on the operations team. After completing all its mission objectives very successfully, *Pathfinder* was extended beyond its nominal thirty-day mission, but the small operations team struggled to keep going after sixty days. We joked about finding the **"do_suicide"** command, though we never found it. After ninety days, though, some element of the spacecraft did fail, most likely due to the severe thermal stress of heating and cooling that the lander was experiencing every Martian day-night cycle. At the time of this writing, the Mars Exploration Rovers have operated almost three times longer than their original mission requirement (90 sols) and show no signs (other than an aging wheel motor) of running out of steam. On September 22, 2004, NASA announced a six-month mission extension for both rovers.

For *Pathfinder,* with the publication of many major scientific papers, Phase E was at last complete in October 1998. And for the people, this whole process had provided a unique training ground for a new generation of engineers and scientists who now had cradle-to-grave experience—that is, from Phases A through E—in one of the most challenging and growing areas of space exploration: landing and operating a machine on another planet.

—BRIAN MUIRHEAD

All it would take was a group of engineers who were committed to perfection yet comfortable with risk, which is a pretty good definition of someone with a bipolar disorder.

It was also a good description of the Pathfinder team.

As Brian Muirhead recalls, "Part of the Pathfinder mystique was the willingness to take risks. To be renegades. Because the Pathfinder group was really per-

On landing day, some of the Pathfinder team assembled in JPL's press conference room. Brian Muirhead is third from the left. COURTESY JPL.

ceived as 'out there.' On the edge. You don't want to get too close to these guys." And, he adds, they liked that reputation. "That's part of what attracted certain members of the group. They weren't afraid to be seen as different or outside the norm."

Brian finds that that quality of the Pathfinder team members actually set them apart from what some people might say is the stereotypical nature of an engineer. "There's a certain group of engineers that is kind of inherently 'nerdy.' And, I think, nerds tends to be introverted, insular. But, for the Pathfinder group, they *had* to be part of a highly interactive team."

The renegade attitude of the Pathfinder group created a different type of team, though. "There was no hierarchy," Brian says, and that made a significant difference in how the team functioned. "Too much structure and a rigid hierarchy can stifle creativity. The Pathfinder team understood there was a chain of command and knew when it needed to be followed. But they also knew and were encouraged, sometimes, to go around it or through it, to get the job done the best possible way."

All the members of the Pathfinder team remember that lack of formal hierarchy—what business managers call a "flat" organization, with only one or two layers between the top and the bottom. They believe it was one of the reasons why they functioned so efficiently. The key Pathfinder leaders—Tony Spear, Richard Cook, John Wellman, and Brian Muirhead, and the new rover leaders—Jake Matijevic and Bill Layman—just didn't care about a formal structure that might get in the way of the work to be done.

Imagine that. Engineers who specialize in the precise analysis of mechanical relationships, chemical interactions, and balance of forces, *not* interested in precise organization flowcharts and pecking orders.

How is such a thing possible?

With Brian, it all goes back to a basement in Chicago when he was seven years old and he built a fort behind the oil tank. Fortunately, under those conditions, the spark that ignited his interest in engineering wasn't literal.

In his basement fort, Brian played with a Kenner Hydro-Dynamic Girder and Panel construction set. "I would build things with plastic girders shaped like I-beams. But what I really liked was a set that had all kinds of fluid systems. They had pumps and you could pump fluid around and open and close valves."

As part of his experiments in "hydrodynamics," Brian recalls he was constantly "making and remaking detergent. Dry soap, mix it with the water, pump it around, dry it out again. I can still remember the smell of that soap. Occasionally, I get a smell of dry laundry soap and it takes me right back to that basement. And it was all behind this big, old quilt that I set up there. For hours, I would rearrange and pump

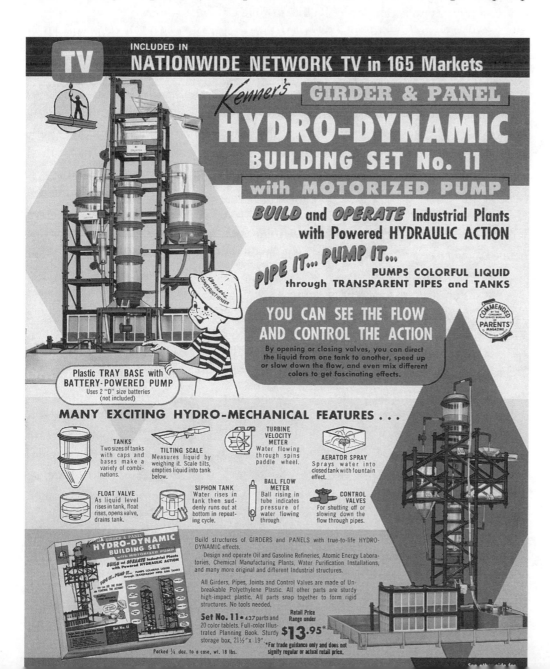

The construction playset that launched Brian Muirhead's career from his basement fort in Chicago to Mars and beyond! COURTESY HASBRO.

all this fluid—it's still in the area of engineering that I love the most. Fluids. Heat transfer. It was intuitive. It was fun for me. I really like it."

For many years, Brian's interest in engineering remained in the realm of play—the word *engineering* hadn't really entered his vocabulary, and in elementary school the level of science education he received was basic. Math, though, was always a subject that came easily to him.

Within his family, both sides were from Kansas farm country. As a kid, Brian spent most summers with his grandparents in Jennings, Kansas. He recalls farmers and their mastery of farm machinery and irrigation as evidence of their being capable engineers in their own right. His closest tie to aviation was his father's uncle, who had been a pilot in World War I, and who then became a businessman at International Harvester.

Fortunately, Brian had access to the bible for young engineers—*Popular Mechanics* magazine. In one issue, Brian says, "I remember reading about the best engineering schools. I remember specifically reading about MIT and Caltech. And I consciously remember thinking: I'd like to go to Caltech someday."

However, upon graduation from Michigan's Bloomfield Hills, Andover High School (near Detroit), Brian didn't go to the University of Michigan as planned or to any other college. Instead, he . . . let's just say, he took some time off. A lot of time.

High school itself wasn't a struggle for Brian. He was a straight-A student through most of it, and in a foreshadowing of his ability to forge a team out of many disparate individuals, he recalls that his high school friends came from every traditional clique, from the jocks to the "nerdniks," and everyone in between. In his final year, in what must surely be an unusual honor for someone destined for a career in engineering, Brian was even voted class clown.

Brian arrived at his high school graduation in cap and gown on his motorcycle—the same motorcycle on which he then took off on a solo journey to eastern Canada. Like many teenagers at that point in their lives, Brian recalls, "I didn't know what I wanted to do, but I knew I didn't want to be an engineer."

After traveling through Canada, Brian visited his grandparents in Kansas and ended up in a construction job, building highways. When the weather turned colder, he managed to put his early "experience" with pumps and fluids to good use when he went to work for a factory that made sugar from beets. Brian was put in charge of the filter section.

"It was kind of interesting, except when the filters got plugged and we had to clean the mud out with sticks and hoses," Brian remembers. But he knew it definitely wasn't what he wanted to do for the rest of his life. "So I went back to Michigan."

For two quarters, Brian attended Justin Morrill College at Michigan State. "What was really neat about it, was this was an experimental college. As freshmen we had full professors teaching every class. So the psych class wasn't Psych 101—it was Psychological Aspects of International Conflict. And physics wasn't Physics 101—it was Great Minds of Physics. Completely different. There was a class that was taught in the cyclotron laboratory, taught by its director. It was really fun stuff."

However, at this point in his life, "fun" wasn't enough to keep Brian's interest, and after two quarters he jumped back on his motorcycle and took off once again.

This time Brian's adventures took him west. He traveled through the Southwest, where he supported himself with more construction jobs and even built mobile trailers. After another winter passed, he traveled to Europe, "Bought a little Porsche in Heidelberg. Toured Europe for nine months and got a job as a motorcycle mechanic in Greece."

When Brian returned to the States, he remembered the appeal of the Southwest and went to Albuquerque. His plan was to attend state college, and since he couldn't afford the tuition as an out-of-state student, he had to establish his residency there first. "So I got a job as a Mercedes mechanic at a BMW dealership. At that point I didn't even know how to open the hood of a Mercedes. But I talked my way into that job. I was mentored by an old German mechanic and when he got homesick, I was on my own. It paid very well. I loved working on cars."

Brian also loved adventure, and the next summer it was time for another one—Brazil. It was a wonderful experience, but for the first time, his spur-of-the-moment decisions about life were beginning to collide with his desire for making something of himself.

When people in Brazil asked him what he did, and he replied that he was a mechanic, Brian realized that he needed to get on with the rest of his life. So Brian asked a friend to send him the catalog for the next semester at the University of New Mexico, and he set his sights on mechanical engineering. He mailed the application from São Paulo.

Brian remembers that the first challenge for the university was to determine his level in math. "There was an exam, but it was funny. I hadn't been in school in three years. I didn't remember anything. But I had spent a little bit of time, just the summer of the year before, studying calculus."

That's an important clue to the nature of an engineer—after being out of school for three years, he spent a summer "studying" calculus on his own. Why? "I always liked solving math problems."

On registration day, back in Albuquerque, "The examiner asked me if I knew any calculus and I said yes. He asked me a couple of derivatives, I knew them, so he put me right into calculus. It was great because I would have been bored to tears below that."

But this time, Brian wasn't bored. Engineering was challenging. He earned his undergraduate degree in mechanical engineering at the head of his class in three and a half years.

Thinking back on those years of traveling and pickup jobs, Brian says, "I learned a lot about myself, that I needed to find a calling that would make a difference. I learned I was a quick study and liked hands-on work. Call me a late bloomer, but I would not trade those three years for anything."

After graduating, the seemingly random threads of Brian's interests, experiences, and education began to knit together. At the University of New Mexico, Brian recalls, "I had one professor who was a genius in the transport sciences: heat transfer, thermodynamics, and fluid dynamics." Exactly the subjects Brian had first explored on his own, in his basement fort, so many years earlier. "I took every one of his classes," Brian says. "His name was Richard Passamaneck. An incredible professor. A wonderful teacher. And he was from someplace called JPL."

Houston, we have liftoff.

At the time, JPL didn't mean much to Brian. "I'd heard of it. When I was a kid, I remember, I did have a set of four books on the planets and rocketry, and I suspect JPL was mentioned in those books. But it certainly hadn't registered."

Now, however, it did. Professor Passamaneck's stories about working at JPL definitely got Brian's attention. Brian recalls, "He strongly advised me against working there. He had been in the propulsion research area, JPL's origins, but as JPL was moving toward being a space systems house, his field of expertise was being downgraded. He no doubt resented it." In spite of his advisor's advice, Brian applied to the Lab for a summer job, and he got it. Not because of his résumé—as Brian says, he didn't have much of one back then. To a place like JPL, 4.0 grade-point averages were a dime a dozen.

So why did Gary Coyle, supervisor of JPL's Configuration and Structure Group, pick Brian Muirhead out of the stack of applications on his desk, without ever even talking to him, let alone interviewing him?

Because of those three years of traveling and pickup jobs that Brian wouldn't trade for anything, specifically his work as a motorcycle and Mercedes mechanic.

At the time, JPL's Configuration and Structure Group was responsible for building the primary and secondary structure on all JPL spacecraft. As Brian explains, "It's a hardware systems group. They are the guys who help design the whole spacecraft and integrate all the pieces. That's one of the things I liked about a mechanical engineering job—you are the integrator. You have to know something about all the pieces you're putting together in order to do that. And that's what really excited me about it."

"I drove my old BMW motorcycle from Albuquerque to L.A. I'd never seen such huge ribbons of flying concrete as the first major freeway interchange coming into L.A. on Interstate 10. When I got to JPL on Monday morning, Gary took one look at me, my full beard and my ponytail, and said, 'I really don't know what we're going to do with you.'"

The year was 1977, and a lot of JPL engineers had been at the Lab since the early '60s, had missed the hippie era, and really did fit the stereotype: short hair, short-sleeve shirts, and more than a few pocket protectors.

It was during his first summer job at JPL that Brian met Bill Layman—the engineer who would become his mentor. Brian worked on a number of different projects, all related to spacecraft design, and conducted computer analysis on punch cards, which, for the benefit of younger readers, was how room-size mainframe computers were programmed in the dark ages before personal computers.

After that summer, Brian would have come back to JPL just to sweep the floors.

But fortunately, he didn't have to resort to such drastic measures because Gary Coyle offered him a real job—working on the just-approved *Galileo* probe to Jupiter.

For the next four years, part of Brian's job involved the ultimate in spacecraft integration—helping integrate the *Galileo* spacecraft to its launch vehicle: the Space Shuttle. *Galileo* was the first interplanetary spacecraft to be launched by the shuttle.

But the shuttle wasn't ready to launch *Galileo* as planned in 1982, and the mission was delayed. The first bump on *Galileo*'s rocky road to Jupiter resulted in a need to slow down the effort at JPL and, having applied and been accepted at Caltech, Brian started his graduate studies in aeronautical engineering. Over the course of a year, during which he remained at his position at JPL, he earned his master's degree.

Not long after that, having completed the delivery of his flight hardware that one day would orbit Jupiter, Brian's office mate told him there was a group supervisor job open. Frank Locatell, one of JPL's greatest engineers—who's probably put more hardware in deep space than anyone—encouraged Brian, saying we need people in leadership roles and you seem to have the right stuff.

As Brian remembers, "My first entrée into real management was in 1984. I became supervisor of a group I really didn't know much about, and the name of the group didn't help much—the Mechanical Support Group." So when Brian got the job, his first act was to change the name to the Advanced Spacecraft Development Group. Remember that—it becomes important later on.

Brian expanded the group from five engineers to twenty. "We had the charter to be a systems group, and to get into new areas, including advanced technology and new missions. That allowed me to attract people from some of the best schools, including MIT, Stanford, University of Texas, and Caltech. I got almost everybody I wanted. I assembled this really awesome group of young generalists—a group that was working on all kinds of wild missions (like a mission to 1,000 astronomical units [about 93 *billion* miles!] to measure the size of the galaxy), using cutting-edge design tools and developing technology for missions like a Mars Rover Sample Return."

In 1986, following the tragic loss of the Space Shuttle *Challenger*, Brian was assigned to work with astronaut Sally Ride at NASA Headquarters to help develop a strategic plan to provide new direction to the agency at that difficult time. His experience with the shuttle helped him lead a team that built the structure and integrated the largest single science instrument designed to be carried up into space and returned by the shuttle. The instrument, a phased-array radar, was called the Spaceborne Imaging Radar Project (SIR-C) (and would fly two more times to generate the best resolution maps of the Earth ever produced).

And though he didn't know it at the time, each assignment at JPL was leading Brian inexorably to Mars.

In physics, they call it chaos theory, that in a seemingly random, chaotic system—like a billowing thundercloud or the arrangement of branches of a tree—there is an underlying order. In fact, an important aspect of the theory is that,

## FBC: WHICH TWO DO THEY REALLY WANT?

*The Birth of Faster, Better, Cheaper*

The implementation of NASA's space missions has always required developers to balance the competing priorities of Cost, Schedule, Reliability, and Performance. In an ideal world, this balancing act would be achieved by a process in which the supplier and the customer for each specific project jointly decide how the different priorities should be ranked in order of importance: Cost-Schedule-Reliability-Performance, or Performance-Reliability-Schedule-Cost, or . . . ?

During the first two decades of the U.S. space program, especially during the race to the Moon, the order of importance was generally Performance first, Reliability second, Schedule third, and Cost a distant fourth. During the 1980s, the order changed so that Performance and Reliability still came first and second, while Cost and Schedule switched places. But still, in both eras, Cost was a secondary consideration and the price tags of missions showed it.

For NASA, the 1980s were dominated by the necessity to get the space shuttle flying, which translated into a very dry period for robotic space exploration. With few missions being flown, the tendency to pile scientific instruments and mission objectives onto the next train leaving the station inevitably drove up costs. This added complexity meant development times grew from a few years to many years (over a decade in the case of the *Galileo* mission to Jupiter, but that was mostly driven by problems with the launch vehicle, the space shuttle).

The uncertainty of funding over such long periods of time often resulted in schedules lengthening even more as the development process slowed down to match the decreasing flow of money. Add to this situation an elevated aversion to risk, compounded by the 1986 *Challenger* accident, and the stage was set for development costs to grow almost without limit. At the same time, ongoing reductions in NASA's funding resulted in a squeeze play that limited the number and type of missions that could even be considered.

Enter, in April 1992, a new NASA Administrator: former aerospace industry executive Daniel Goldin. While some considered him a controversial choice, no one would disagree that Dan Goldin loved NASA and the exploration of space, and that he had a vision for both. Put succinctly, he wanted to "darken the sky with spacecraft." To do that, he picked up the just-emerging idea of "Faster, Better, Cheaper," known in NASA parlance as FBC, and ran with it.

The Discovery Program, a new space initiative conceived by NASA's then Associate Administrator for Space Science Dr. Wesley Huntress, became the testing ground for FBC, and the Mars Pathfinder mission became its poster child. The rules for this program were simple. Each Discovery mission had to be *Faster*—development time was limited to three years; *Cheaper*—the total spacecraft development cost was limited to $150 million in 1992 dollars (not including launch vehicle or operations) or about $200 million in 2004 dollars; and *Better*—missions would be novel, science driven, built around specific, precisely defined science objectives that would not be allowed to expand during development.

When the idea of FBC was first discussed at JPL, the common reaction was "Faster, better, cheaper? Which two out of three do they want?" But in the case of Mars Pathfinder, NASA Headquarters made it very clear to the mission's personnel that two out of three was not acceptable, particularly if the priority that wasn't met was Cost. As for Schedule, the launch date was fixed by the geometry of the planets (and you only get a shot at Mars every twenty-six months). As for Performance, the mission objectives were clear: land an operational spacecraft on the surface of Mars, spending about the same amount of money James Cameron used to make his movie *Titanic*—a lot by Hollywood standards, minuscule in the realm of space exploration. So the order of priorities was now inverted: Cost, Schedule, Reliability, Performance.

Between 1992 and 1998, NASA conceived and launched ten FBC missions, with only one failure. (The *Lewis* Earth-observation satellite, a technology-demonstration mission, was lost in Earth orbit due to an operations failure.) In addition, a number of Department of Defense missions were undertaken and generally successful. Overall, the FBC success rate was 90 percent, which is very good for "high-risk" missions.

But by the end of the decade, that 10 percent failure rate suddenly jumped with the near-simultaneous loss of two sister Mars missions, and the halcyon era of FBC appeared doomed (and for the critics of FBC, it was "I told you so" time).

—B.M.

| DEVELOPMENT COSTS (NOT INCLUDING LAUNCH VEHICLE AND MISSION OPERATIONS) IN 2004 DOLLARS (MILLIONS) | |
| --- | --- |
| *Mariner* '62 | $539 |
| *Mariner* '71 | $606 |
| *Viking* '76 | $3,931 |
| *Voyager* '77 | $956 |
| *Galileo* '89 | $2,097 |
| Mars *Pathfinder* '96 | $232 |
| *Cassini* '98 | $2,514 |
| *Deep Space 1* '98 | $114 |
| *Mars* '98 | $225 |
| Mars *Odyssey* '01 | $322 |
| Mars Exploration Rovers '03 | $615 |
| Mars Science Laboratory '09 | $1,320 |

though it might seem impossible to look ahead and say where a particular set of starting conditions might lead, in retrospect it is sometimes possible to look back and see the underlying pattern. The same could be said of human careers.

So it was that out of the chaos of his interests, his life, and his career choices, Brian came to work on Wes Huntress's Discovery mission to put a rover on Mars.

## RESISTANCE IS FUTILE

By 1992, the concept for a low-cost Mars surface mission called MESUR had been studied and refined by Tony Spear at JPL for almost two years, and it was still just an intellectual exercise. In NASA's acronymical parlance, the name stands for Mars Environmental Survey. Okay, so it's not exactly an acronym.

As originally developed by NASA's Ames Research Center in the heart of California's Silicon Valley, the big-picture idea for MESUR was to create a planetary network of about sixteen small automated surface stations widely scattered around Mars, in order to study the planet's internal structure, weather, and local surface conditions, with all stations connected by Mars-orbiting communications satellites. In a way, it shared some features with the Discovery program, especially the idea that by sending sixteen landers to Mars, the loss of one or two wouldn't doom the entire program.

At about the same time, the Discovery program itself had found a supporter in NASA's new administrator, Daniel Goldin, appointed by President George H. W. Bush on April 1, 1992.

Dan Goldin had come from the aerospace industry, and he became the champion within NASA of the business concept known as "Faster, better, cheaper." Dan was equally driven to increase the tempo of NASA's launches, and he saw a perfect fit between Wes Huntress's low-cost Discovery program idea and his own new direction for NASA.

For Brian Muirhead, though, all these developments were little more than background noise to his work on the SIR-C mission. He was now manager of the mechanical systems section, responsible for 120 people working on everything from a new camera for the Hubble Space Telescope (the Wide Field and Planetary Camera 2—which would have in its optics a fix that would clear the vision of the Hubble and open its window onto the universe) to the new *Cassini* mission to Saturn. He also had two lovely daughters, ages four and two, and, for better or for worse, a marriage that had ended on amicable terms. As Brian puts it, professionally and personally, "My hands were full."

But Tony Spear had other ideas.

After the two years he had spent refining MESUR, it was clear that there was not enough money to implement the full MESUR mission. But MESUR had a completely reasonable fallback position—a so-called pathfinder mission in which a single spacecraft would be sent to Mars to demonstrate the technology for the fifteen others to follow.

## NAVIGATING THROUGH JPL

*Careers in High Tech Exploration*

One of the truly remarkable things about JPL is the amount of responsibility you're given right from the beginning in your job. My first assignment at JPL was as a lead engineer (we call them cognizant engineers) for the structure that would carry the radio-isotope power source on the *Galileo* mission to Jupiter.

So here's a guy, right out of college, responsible for the design, manufacturing, test, and delivery of mission-critical hardware to the planet Jupiter. In addition I had two other jobs that allowed me to travel and work with other NASA centers and industry. The most notable was the Johnson Space Center, which is responsible for the design and operation of the space shuttle, and the Kennedy Space Center, responsible for launching the shuttle. I also worked with the Department of Energy in Washington, D.C., and General Electric in Valley Forge, Pennsylvania, responsible for building the new power source for deep space. I remember, as just a junior engineer, interacting with shuttle managers, including Glenn Lunney, one of the engineers who helped get *Apollo 13* back safely. I felt I was truly at the center of the space program—and I was.

Of the approximately 5,400 people who work at JPL in 2004, about 3,800 are scientists and engineers supported by 1,600 people in business, manufacturing, security, and other key support roles. The management structure is a classic matrix structure divided into the "line" organization and individual projects. The project side is mainly made up of managers, lead scientists, and system engineers who direct a project. The line side is made up of managers and the lion's share of the working engineers and scientists.

The matrix works more or less this way. As a project moves from concept to development, more and more people are assigned out of the line to support a specific project. As they complete their jobs, they move off one project on to another, or on to a study or technology task. Although there is no place for people to wait until their next job, JPL managers are remarkably good at phasing people in and out of jobs, generally without any loss of staff. (This may be due in part to a combination of talent, versatility, and a willingness to apply people outside their areas of expertise.) The project leadership and a core team tend to stay with the project from beginning to end.

So, what might someone's career path look like at a place like JPL? Let's start with the linear model where you come in at the bottom, work your way up the technical side, and then move into management, either project or line, right up to project or division manager.

The speed at which you move through this process could be relatively constant. It used to be there was only one way you got to the real benefits (like the senior-staff retirement plans and the greatest prize of all, on-Lab parking!), and that was to become a manager. Not too surprisingly, that's what everyone aspired to become, which often resulted in the loss of really good technical talent and the creation of mediocre and/or unhappy managers. But today, the current system has two senior technical job classifications that enjoy the same perks as managers but remain entirely technical, providing a true dual-path career ladder.

One of the best things about JPL is its acceptance of creative, nonlinear behavior. I've often called JPL a glorified graduate school, for its casual air, tolerance of individuality, and the sheer brilliance and drive of the staff. As in many other organizations, the real rate of advancement and direction (or "velocity") of a career is often determined by opportunities that arise because of accident, of luck, or because of who you know and who knows you. But at JPL career velocity is also a strong function of your raw ability, and the Lab is particularly good at giving people opportunities ahead of where their résumés say they should be.

For example, the Mars Pathfinder team peaked at about three hundred people, and if you could get them into one room and look around, you'd have seen the average age was in the low thirties, with only a few gray beards and bald heads thrown in. Because most of these people were doing a new job—a job they hadn't done before or had never been done at JPL—there was a significant element of stretch for nearly everyone. And that made it fun and more than just a little bit scary. It also was a great growth experience and out of the crucible of *Pathfinder* came many of the leaders of the very successful Deep Space 1 mission and current Mars landers: the Mars Exploration Rovers.

For those interested in leadership positions, JPL offers many opportunities. Leadership roles can begin right away as a cognizant engineer. In this position, you must lead a small group of designers, builders, and testers. You'll have a budget, schedule, and responsibility for the delivery of a piece (or pieces) of hardware or software. Most big aerospace companies would break this role up into separate functions and not give a single individual this kind of hands-on, end-to-end responsibility. JPL's size and the one-of-a-kind nature of what we do allows and actually requires this approach. So out of this single job comes someone with development, leadership, and some systems-engineering experience—all valuable currency for moving up the ladder.

Ideally, after a couple of cycles as a cognizant engineer, you're ready to move up. The next rung is achieved by becoming either a group supervisor on the line side, or a project element manager (PEM) of an engineering subsystem. Both jobs require and develop technical and people management skills. A group supervisor could lead a group of five to twenty-five workers, and a PEM could be managing a task whose budget is ideally between $5 million to $20 million.

The next step could be to lead a section of about one hundred people, or take on a bigger subsystem job up to $20 million to $50 million. Advancing up the line side would lead to becoming a division manager responsible for around five hundred people, or a flight systems manager assigned to an in-house or contracted job. Given the large number of

All 177 acres of the Jet Propulsion Laboratory in Pasadena, California, home to about 5,500 workers. COURTESY JPL.

And, of course, all this great engineering wouldn't be possible without the support of our business, facilities, outreach, and technician staff, who keep the money flowing, put hundreds of millions of dollars on contract with industry, build and test our hardware, maintain security, and keep our computer systems running.

JPL is a small town covering 177 acres that's open 24/7, with its own fire department, medical staff, three cafeterias, a company store, and a museum. We're a close-knit community sharing a common set of goals and values, with enormous enthusiasm for our work, and we have a lot of fun doing it . . . at least most of the time.

Many of us came to JPL with the thought that we'd work there for a while, then try something new; maybe even get rich at some profit-making company. Some of our alumni have done just that, like Dennis Tito, the first paying tourist in space, though he made his fortune on Wall Street, not in aerospace. But many of us have never left, and those who leave often come back.

JPL's not perfect, but then, what organization—that's been around for almost fifty years, has over five thousand employees, operates under the scrutiny of the media, works for the government, and thrives on change and new challenges—is? We sometimes talk about succeeding in spite of the institution, or we complain about the workload, and there are some very long days—forty minutes longer in the case of Mars missions.

But all things considered, the Jet Propulsion Lab is one of the best places to work, on this planet or anywhere else. And I believe that my colleagues at the other NASA centers and the big and small companies of the aerospace industry, without whom the exploration of deep space would not be possible, feel the same way about the people and organizations they work for.

—B.M.

projects of different sizes, one could even land a project manager (PM) or deputy PM job. In my opinion, the best job at JPL is to be a subsystem lead or flight systems manager for a major in-house project. You get to work a hands-on job, building challenging and unique spacecraft with the best of the best. For an engineer who loves solving tough problems, likes working with people, and isn't afraid to put him/herself on the line, it just doesn't get any better than that.

The final level, just below the director of the Laboratory, is the executive council made up of program and directorate officers (responsible for groups of projects) and the leaders of the business, mission assurance, and line-management organizations. This is where the policies that guide the Lab's development and implementation strategies are made and monitored. This is also where key

senior staffing decisions (like project manager) are made.

Besides providing great career opportunities for hands-on engineering and leadership, JPL produces some of the very best spacecraft systems engineers on the planet, largely because we don't keep people in restricted "silos" of technical expertise, we give them lots of opportunities for growth. Systems engineers are by definition generalists, big-picture people with knowledge and experience in a wide range of fields. They are the glue that hold a project together, and the grease for the processes (especially decision making) that are needed to keep things moving. They are the architects of new missions and new designs, the people who naturally fill the vacuums, and the firefighters who solve the interdisciplinary problems that always show up in exotic deep-space missions.

Thus, for MESUR to go forward, it wasn't necessary for NASA to sign off on the whole sixteen-lander package at once.

But the Pathfinder component could possibly be done—as a Discovery mission changing MESUR Pathfinder into Mars Pathfinder. So, as the proposed mission transferred into the Flight Project Office, it was time to get serious about staffing, even though an official go-ahead would be dependent on NASA's new budget. Would there be a new line item for Discovery or not?

Toward the end of 1992, Tony met with Brian and told him he was going to advertise for the position of Pathfinder flight systems manager. On the surface, it was going to be an impossible job. Whoever was crazy enough to take it on would be responsible for the design and building of the entire spacecraft, *and* for coming up with a way to do it on the ridiculously low budget in an almost as ridiculously short schedule.

Oh, and Tony thought Brian was the guy who could pull it off.

Brian actually made a list labeled "pros" and "cons." On the one side of the list, he had a stable job and a solid career path. On the other side, he had an incredibly risky mission that had no guarantee of even being funded. As Brian recalls, "If I'd still been married, I probably wouldn't have even considered *Pathfinder*. It would be crazy to leave a sure job for something that may never happen."

Still, Brian checked out Pathfinder with his friends and mentors at the Lab. Everyone was encouraging, but no one was optimistic. "The assignment looked impossible, the ability to succeed remote at best." Even JPL's future director, Charles Elachi, counseled him to wait until the NASA budget came out to see if NASA was serious.

Given all that, how could Brian resist?

He couldn't.

In April 1993, Brian Muirhead became flight systems manager for *Pathfinder*.


## SETTING THE STAGE

Flight systems manager. Project manager. Flight operations manager. Software lead. What do all these titles mean?

Brian explains. "On a typical mission, there's a core leadership team of the project manager, project scientist, flight systems manager, mission manager, and the science manager.

"Flight systems managers are a unique group of people. It's a short list. But the job still tends to be, in big projects, heavily focused on the management of resources and budgets. You delegate the responsibility for the design work to a lead spacecraft systems engineer—the guy who makes most of the calls on the system design."

Which brings us to another point at which the Pathfinder mission veered off from business as usual. On Pathfinder, Brian filled *both* positions. "I was the flight systems manager *and* the lead system engineer, which meant I made most of the system calls." But Brian is quick to point out that: "I didn't reason them out brilliantly all by myself. I formed a triumvirate with Rob Manning and Curt Cleven, two great engineers whose technical skills complemented my own. And I had other key "guys," like Guy Beutelchies, Chris Salvo, and Cathy Cagle, as well as subsystem leaders Dave Lehman, Carl Buck, Leslie Livesay, John Klein, and Jay Dettinger, who were the inner circle that collectively did a lot of the system engineering.

"But ultimately there's one guy who says, 'Okay, that's the way we're going.' And I had reasoned that because of the budget, schedule, and technical challenges we faced, we were going to have to balance the resources—meaning dollars and the system engineering resources like mass and power—with every decision we would have to make. We couldn't make any decision without consideration of all those factors."

In the usual way spacecraft missions were organized according to a strict hierarchy, that balancing act didn't play out at the level at which the engineering decisions were being made. "Typically," Brian explains, "if you're the systems guy, you worry about the mass and the power and the system. But you never worry about the budget, and you never worry about the schedule." As Brian saw it, "That's been one of the reasons that past projects tend to overrun their budgets. The guys that are making the great technical decisions are spending money and schedule like crazy, and the project management is trying to shovel money as fast as they can to fill this void. But in the new 'faster, better, cheaper,' paradigm of a Discovery mission the budget and schedule are fixed. So, it seemed essential to me that one person was going to have to be at the nexus of all that decision making if we were to have any chance of meeting the challenge."

Under this plan, Brian had two key meetings each week. "One was the flight systems *management* team, which usually lasted an hour, and the other was the flight system *implementation* team, which was the engineering team.

"I led both of those meetings, and remained hands-on in the middle of the process the whole way. It wasn't like I was brilliant, but I did know where to go to get the brilliance."

As part of Brian's approach to management, he not only shared two jobs with himself, he also made certain that every member of the team shared his sense of priority about managing two of the most critical commodities affecting mission design and success—mass and money. Mass, because there was a strict limit on how heavy the *Pathfinder* spacecraft could be; not so much at launch but definitely at the time of entry at Mars. Money was equally critical because Pathfinder was operating under an unbreakable cost cap—something that hadn't been strictly applied in the past.

It was common for projects to get into trouble with unanticipated technical problems or underestimating the number of people required, resulting in budgets being used up ahead of schedule, leading to cost overruns. At the time of Pathfinder, the experience throughout the aerospace industry—and JPL was no exception—had been very poor performance against original budget estimates. Stories about $500 toilet seats, unfortunately, didn't surprise anyone in the industry.

But Pathfinder, as a Discovery mission, was going to change all that, or literally die trying.

Everyone on the team felt that the instant it even *looked* as if Pathfinder was going to exceed its budget, the mission would be up for cancellation. There would be no last-minute cost overruns, no deathbed reprieve. Each dollar was as fixed a resource as each ounce of the spacecraft.

The idea of building a team with this level of sensitivity wasn't something that had just come to Brian. He had been working to find ways to deliver subsystems under tight budgets before, and just as Wes Huntress had found a champion in Dan Goldin, Brian's approach to creative management fit perfectly with the requirements of a faster, better, cheaper Discovery mission.

A key element of his management method involved a certain type of engineer that he had been encouraging when he was a group supervisor. "The role of my group was to grow a new crop of generalists, top-notch systems people. And the best way to do that is by hands-on work.

"If you're going to be a systems engineer and you're going to design a spacecraft, you need to know something about every aspect of the system: command and data handling, power, attitude control, propulsion, temperature control, structures—and don't forget mission design and operations. So I put people in roles where they'd have to think about these subsystems, do design work and calculations involving their performance, and develop integrated designs. I wasn't trying to put other specialist groups out of business. I was trying to train a generation of generalists the only way I knew how."

For the first six years that Brian worked at JPL, he had never seen a budget, because at his level that information wasn't necessary for him to do his job. But for Pathfinder, "Part of what we had to do was get everyone aware of budget considerations, even at the very lowest level. And that meant, in many cases, that cognizant engineers [the name given engineers responsible for a specific hardware or software "deliverable"], had to develop their own budgets and schedules.

"They weren't necessarily good at it," Brian admits. "Some of the people were notoriously bad at it. But everyone developed a sense of ownership of the budget. However, just doing the budget isn't enough to own it. You have to take it seriously. So we made cost consciousness part of our project culture. We made it clear that everyone needed to think dollars in every decision that they made. But that didn't mean you made dumb decisions—mission success was still paramount. We had to be sure we weren't being penny-wise but pound-foolish. But as I've seen in the past, when you give a group of people a tough but clear challenge they will rise to it, every time."

Brian is certain that part of the reason why Pathfinder met its cost constraint is because, "We got that message out right from the beginning, across the entire project, and we held to it. The project leadership team, and especially Tony Spear, deserve a lot of credit for that—everyone was watching every expenditure." Brian has to smile as he adds, "But since everyone knew that the flight system was where the biggest risks were, that's where the budget reserves would go. And I didn't disappoint. Though Tony made sure I defended every new cost increase to my peers and only then did I get the money. But I can say with a clear conscience, not a penny was wasted."

## CHARLIE BROWN, LUCY, AND THE FOOTBALL

So far, this has been a pretty one-sided account of going to Mars. Mostly, we've talked to and about engineers and managers. After all, they're the people who design and build the spacecraft that carry scientific instruments into orbit and to other worlds.

But keeping engineers happily employed is not really what going to Mars is all about.

It's Science.

What's the difference between an engineer and a scientist?

Theodore von Kármán, the founder of JPL, has one definition. He said: "Scientists discover what is, engineers build what has never been!"

A joke that's made the rounds of engineers offers another definition: The graduate with a science degree asks, "Why does it work?" The graduate with an engineering degree asks, "How does it work?" The graduate with an accounting degree asks, "How much will it cost?" And the graduate with an arts degree asks, "Do you want fries with that?"

With apologies to arts graduates, the difference between "how" and "why" is a good thumbnail description of the difference between an engineer and a scientist. But we must always remember that there is a considerable overlap in science and engineering, especially in the exploration of space. Scientists and engineers must understand each other's worlds and priorities and find a balance between them. Scientists set the goals and specify the environments that engineers build to. Engineers build the instruments that make the measurements that scientists base their discoveries on. The best experimental scientists are (or become) pretty good engineers—Dr. Steven Squyres is a good example, and we'll meet him soon. And many engineers do what they do because they love science. Although the relationship can sometimes be difficult, it is always symbiotic.

With that in mind, let's take another look back at the beginning stages of the Pathfinder mission to Mars.

As originally defined, Pathfinder was to be a "technology demonstration" mission. The sum total of its original science objectives—make that, objective—was to take a picture. Once the Discovery concept had proven itself, science would drive all the subsequent missions.

So how was it that five years after Brian Muirhead signed aboard a technology demonstration mission, that this mission became one of the most productive science missions to another planet that NASA had ever launched?

Blame Dr. Matthew Golombek, Pathfinder's project scientist—project scientist for a project that wasn't supposed to have any science.

You might not remember Matt's name, but if you've ever seen any news coverage of the Pathfinder or MER missions to Mars, you know his infectious energy and smile, and you know his distinctive laugh. Quite possibly, these days Matt is the happiest scientist in the world, and his enthusiasm and passion for his work is contagious.

But unlike many of the people on the Pathfinder and MER teams, Matt can't recall how his journey to Mars started. As he remembers his early childhood, "I was always interested in how things work and why they are the way they are. I think it's in your genes."

Matt doesn't think engineers and scientists are alone in this aspect of their personalities. "I think it's true for everyone—you like doing what you do because there's something innately interesting in it to you and you don't know why that is—it just is."

One of Matt's chief interests was geology, and he joined the geology club in his high school. However, just like many of his colleagues at JPL, while in high school he took many courses, including calculus, advanced physics, and advanced chemistry "just because they were interesting." Sort of reminds you of a certain engineer on a road trip who studied calculus on his own one summer, just for fun.

When it was time to think of college, geology remained Matt's primary scientific interest, "if only because I liked being outdoors. I liked hiking and backpacking and stuff like that, and for geology you do fieldwork outside.

"I remember always being interested in the physical layout of the Earth's surface—why there were mountains, and why there were valleys, and how did they get to be that way. And that, of course, is what one branch of geology is trying to figure out. In some sense, I guess, I connect with the Earth."

Fortunately, being one of the foremost specialists in the geology of Mars doesn't prevent Matt from going on field trips—something that won't be possible on the Red Planet for years to come.

"For the *Pathfinder* landing site evaluation," Matt recalls, "we actually went to a potential Earth analog." An "analog" is a specific place on Earth that's believed to have geological or environmental conditions similar to a particular place on Mars. For *Pathfinder*, the analog to the potential Mars landing site was in the Channeled Scablands of Washington state. "Not only did we lead a field trip with many of the project members, but a lot of the EDL engineers and people who were worried about the landing site came." Matt adds with his trademark laugh, "And, oh, did their eyes get wide!"

Matt's referring to one of the first locations he took the team—very rocky, very dangerous. "We took them to this place and we told them that this is the worst that it could ever be, and they were so scared, they were ready to call off the whole thing! So we made a special stop the next morning to specifically take them to a place we thought was more realistic, with fewer rocks, and I remember them going around saying, 'Oh, yeah, this is no problem.' They even said, 'You know, these rocks aren't as sharp as we thought they were going to be, so we think this is going to be fine.'"

But years before Mars appeared on his personal horizon, Matt remained focused on Earth geology, though when he entered his undergraduate program at Rutgers College, he still wasn't 100 percent convinced that was the field that would become his career.

Matt allows that in his first year of college, "I probably spent more time partying than anything else." But he also took a range of diverse classes, from the History of Religion, and History of Western Art, to philosophy. But midway through his sophomore year, he was required to declare a major, so Matt paid a visit to the chair of the geology department. The meeting couldn't have gone better.

"The chair was Ray Murray, and he was in a faculty meeting. But when the secretary told him that there was a prospective student out front, he kicked the faculty out and ushered me in—I was so surprised! Then he sat and talked to me for an hour about what the requirements were for a geology major, and I was very impressed with that personal interest."

Almost by necessity, Matt's study of geology concentrated on Earth. "I remember taking an undergraduate course on the geology of the Moon and planets, but there wasn't much on the planets back then."

The Moon, however, was a different matter, and its geology had been studied relatively extensively by lunar orbiters, and by the investigations conducted by Apollo astronauts. Thus, as a graduate student at the University of Massachusetts at Amherst, "I wound up doing a master's degree in structural geology on the Moon. That was my first foray into planetary."

But then practical concerns intervened. As interesting as lunar and planetary geology were, Matt decided there might not be too many job opportunities in that field. So, for his PhD, he returned to Earth and earned a "terrestrial" degree.

However, planetary geology continued to intrigue him, and he got his first real taste of it—and Mars—as a graduate student.

In 1976, NASA had successfully placed two *Viking* orbiters into orbit of Mars, and two *Viking* landers on the Red Planet's surface. A primary mission objective for all four spacecraft was an operational lifetime of ninety days. But when NASA builds for success, it doesn't kid around: Orbiter 2 operated for just over two years, Lander 2 for just under four years, Orbiter 1 for over four years, and Lander 2 for more than six!

Of course, whenever mission hardware continues to operate past its original schedule, NASA has to find the money and personnel to continue to command the spacecraft and receive and process the continuing flow of data. At this stage in many missions, NASA welcomes proposals from scientists who were not necessarily part of the original science team, inviting them to suggest new investigations the spacecraft might perform.

One of the scientists Matt Golombek worked with as a graduate student had received a grant to work with the *Viking* orbiters for the Extended Viking Mission in 1977. Matt became his student assistant and made his first trip to JPL as a Viking guest investigator.

He remembers that by then, the Viking project office "was this small little place with about ten or twelve people. Basically, the engineers were looking for places to take pictures with the orbiter." There were a few areas on Mars that Matt was interested in, so he suggested them to the Viking imaging team. "And you know, a couple of days later pictures came down. It was really cool."

After graduation in 1983, Matt Golombek came to work at JPL full time, functioning as both a terrestrial and a planetary geologist. By 1987, Mars had become one of his specialties. But, unfortunately, at the time, that specialty gave him little to smile about.

At the time, NASA was still in its *Battlestar Galactica* days, and every Mars mission Matt studied and worked on grew larger and more complex until it collapsed under its own budgetary weight.

One of the first studies he worked on was the Holy Grail of Mars mission planning—a Mars Sample Return mission. Very simply, an MSR involves a vehicle landing on Mars, scooping up a few ounces of Martian rocks and soil, loading it onto a Mars Ascent Vehicle—called, you guessed it, an MAV—which then launches into Mars orbit where an Earth Return Vehicle docks with it, takes on the Mars sample container, then flies back to Earth. And keep in mind this is the simple version.

Well, by the time the engineering and scientific bureaucracies had their way with those plans—adding, among other things, the equivalent of a Hubble Space Telescope that would be placed in orbit of Mars to locate prime sample-gathering sites—the budget had hit a mind-boggling $10 billion. As Matt describes it, the mission "just kept getting bigger and bigger and more and more absurd until it was canceled."

Then there was the Presidential Space Exploration Initiative of 1989. Unlike NASA's current Vision for Space Exploration, which was jointly developed between political leaders and NASA, the 1989 initiative was solely a political proposal. NASA was suddenly given the task of planning the human exploration of Mars and told to come back with a schedule and a price tag.

After NASA's collective jaw had stopped dropping, every part of the agency had its say, adding so many bells and whistles to cover every conceivable eventuality that the price the agency came back with, adjusted for inflation, came to $600 *billion*.

There's a technical term for $600-billion space programs: dead on arrival.

Was Matt discouraged by working so long and so hard on proposal after proposal that inevitably got shot down? Yes, but he also had a philosophical approach to the situation.

As Matt Golombek sees it, "Scientists are like Charlie Brown and Lucy and the football. I bet Charlie Brown knew that every time he would run up to kick that football, she was going to pull it away. But he always had some hope." Not exactly Sartre, but definitely one that fits.

And that's exactly what Matt and the NASA science community had. "A little bit of hope that we actually would go to Mars and learn something."

So every time *Battlestar Galactica* hit the scientists' desks, "It was an opportunity to go to Mars, and that was great. But we always had this feeling it isn't really going to happen."

In the tradition of all great stories, it's just at the point when the hero is about to be defeated that a new opportunity arises. And that's exactly what happened with Matt.

The day came when the Mars study that hit his desk was Tony Spear's MESUR—the Mars Environmental Survey program.

As originally conceived, the full, sixteen-lander mission concept didn't have a hope of ever being funded. But a single-lander "pathfinder" component would cost no more than . . . oh, about what Wes Huntress had set as the budget for his Discovery missions.

Can you see the puzzle pieces coming together?

Wes Huntress had a program and a goal to land on Mars.

Tony Spear was convinced he could accomplish that goal for the money.

And Matt Golombek had a Mars mission that needed just what Wes and Tony had to offer.

However, of all the different puzzle pieces that were coming together so smoothly (and unexpectedly), there was one piece that didn't quite fit.

It was Matt.

After all, Mars Pathfinder—as the Discovery mission to Mars was now known— was a *technology* demonstrator, and Matt was a scientist. In fact, Matt still wonders if he was assigned to the project because certain managers at JPL thought the whole idea of building and launching a Mars lander for $150 million "was so stupid, it wouldn't matter if I screwed up. It didn't have any science anyway. It was strictly an entry, descent, and landing mission. It'll keep Matt out of our hair."

If you're detecting a sense that scientists and engineers look at missions from different points of view, you're right. It's a time-honored tradition, no different in its way from the relationships between architects and builders, and even between movie scriptwriters and directors.

At JPL, however, this relationship between scientists and engineers is a necessary part of the job at hand, and as such, the relationship has been institutionalized.

"Every mission has a chief scientist," Matt explains. "The job of that scientist actually changes dramatically over a period. In the beginning, you're the conduit between the actual engineers who build the spacecraft, who generally don't know squat about science," he says bluntly, "and the science community, which is, presumably, the end user of this mission." Matt is quick to point out that he doesn't fault the engineers for not concentrating on science. It's not their job, any more than it's a scientist's job to focus on engineering.

But that makes the role of the project scientist even more crucial. "You need to have people who're watching closely during development of the mission. And it's generally the two people who head the mission," he says. In the case of Pathfinder, those two people were Project Manager Tony Spear and Project

## FBC: THE NEXT GENERATION

*The Legacy of Faster, Better, Cheaper*

Up until December 1999, the success rate of FBC missions was 9 out of 10. Very good shooting by space system development standards, especially when that success rate applied to both accomplishing the mission and staying on budget. For *Pathfinder,* some of us joked that we should have gotten at least half the credit for mission success for just making it to the launch pad.

Before, during, and after *Pathfinder,* there were a number of groups who sought to define the Faster, Better, Cheaper principle. But nothing captures the underlying character or the paradoxical nature of FBC better than the words Dan Goldin said to me as we walked into the clean room at JPL to see the *Pathfinder* spacecraft as it was just starting to be assembled: "I want you to take risk, [pause for effect], but do not fail." The essence of FBC is the resolution of this riddle. And the threefold answer is: Be creative about how you do your job (intentionally do things differently); be innovative and use new technology; and most important, put together a talented, trusted team who own their jobs (and get out of their way!).

Of course, NASA did not invent FBC, but rather adapted it to work within NASA's open, high-visibility environment. FBC principles originated inside places like the famous Lockheed Skunk Works, under Kelly Johnson, which produces amazing results (like the SR-71 Blackbird) behind the black curtain of Defense Department secrecy. These clandestine programs and the *Clementine* mission, under Colonel Peter Rustan, who wasn't afraid of taking risks, were progenitors that we at NASA studied extensively.

The FBC spirit of innovation established a number of new ways of doing business within NASA:

- Acceptance of an absolutely unchangeable total budget by every member of the team

- Limited scope, clearly defined and adhered to

- "Ownership" by every member of the team of their part of the job—i.e., delivering needed functionality at the highest quality

- Teamwork over paperwork

- Extensive use of peer review with limited amount of formal external oversight

- Good communication throughout the project, enhanced by having as many team members as possible work in the same location

- Hands-on leadership: microknowledge, not micromanagement

- Fast, effective decision making

- An atmosphere of openness and honesty across the entire project

- Development of robust systems that are tolerant to what engineers call the "unknown unknowns"—that is, all the things that could go wrong or be significantly different than expected

- Use of new technology, focused and proven in the heat of the project

- Thorough testing. Never saying: "That's close enough"

Once the concept for lower-cost missions took hold, the NASA administrator's passion for cost cutting was extended everywhere. While the Discovery Program had clear guidelines, others, including the developing Mars Program did not. Before *Pathfinder* was off the ground, or even much of its technology demonstrated in the laboratory, let alone on Mars, the Mars '98 missions were proposed and approved for the next Mars launch opportunity in 1998 (twenty-six months after *Pathfinder*).

But the subsequent loss of those two Mars '98 missions, Mars Climate Orbiter and Mars Polar Lander, put the expansion of FBC into a very high-g dive. The causes were widely debated and Tom Young (a twelve-year NASA veteran and former president and chief operating officer of Martin Marietta Corporation) headed the independent review committee that made the official assessment of what went wrong.

In short, the root cause of the loss of the Mars Climate Orbiter was due to an embarrassingly simple conversion error between metric and imperial units (a conversion between pounds and newtons equal to 4.45). The discrepancy in the data caused by this magnitude of an error (this error is of the same order as a decimal-point mistake, which happens easily and is easily recognized) was recognized but not corrected due to communications problems (and a lack of formal documentation of the discrepancy) between the builders (Lockheed Martin Aerospace) and the operators (JPL). The Mars Polar Lander was lost due to a software design flaw that should have been caught in the testing phase but wasn't. Both losses were laid squarely at the feet of Faster, Better, Cheaper. In hindsight, though, a number of the principles listed above were not followed. But that's another sidebar.

Few would disagree that Mars '98 was pushing the envelope much too hard—the mission called for building two spacecraft for the price of one *Pathfinder.* But they both made it to the launchpad on time and on budget and, based on the quality of the flight system development work at least the Mars Climate Orbiter should be orbiting Mars today. It is entirely possible that if the MCO had achieved orbit and met its mission objectives, Faster, Better, Cheaper would have

NEAR—the Near Earth Asteroid Rendezvous mission—was the first Faster, Better, Cheaper project to come out of Wes Huntress's Discovery program. The NEAR *Shoemaker* probe rendezvoused with the asteroid Eros, studied it from orbit for a year, and then, in a delicate feat of navigation, had its orbit slowly lowered until it "landed" on the asteroid on February 12, 2001. Though the probe was designed to function only as an orbiter, it returned information from the asteroid's surface for another two weeks. COURTESY NASA/JOHNS HOPKINS UNIVERSITY APPLIED PHYSICS LABORATORY.

continued to be a guiding principle at NASA, at least for a while. However, given the tendency to keep pushing the envelope, especially on cost, it was clear that sooner or later failure was inevitable.

One of the things that JPL did very well after the '98 failures was to capture and institutionalize a set of principles and practices to guide all missions in the future. Associate Director Tom Gavin and systems engineer extraordinaire Matt Landano (currently head of Mission Assurance at JPL) had the vision and expertise to tap into the core of what made JPL successful and create two documents capturing that knowledge (and keep it current). The documents are referred to as The JPL Design Principles and The Flight Project Practices. These are not hard-and-fast rules (although there's a tendency to look at them that way) but things every project team needs to think about and decide consciously whether to deviate from them or not, depending on the mission requirements. Balancing the use of best practices and innovation starts with knowing the best practices. Carly Fiorina, CEO of Hewlett-Packard, had an ad that ran a few years ago in which she said of her direction for the company: "Keep the best, invent the rest." That sums up my view of FBC pretty well.

FBC clearly broke the mold in a number of areas, especially in the use of new and innovative technology. Before FBC, the first question asked of anyone proposing a particular design was: "Has it flown before?" If not, forget it. But FBC showed that new technology was critical to meeting tight cost constraints and could be flown successfully if the application was sound, the design robust, and the testing rigorous.

Many of the FBC principles are actually part of the normal way of doing business today. But the level of oversight is much higher, and the willingness to take risk is much lower, especially for Mars missions like MER. The net effect is more expensive missions. However, there are still missions and science instruments, in particular the smaller, lower-visibility missions, where either the stakes for failure are perceived as acceptable, or the requirement to meet a strict budget is still paramount. These missions are still doing business faster, better, cheaper.

For the people who have worked on the FBC missions, the sense of camaraderie, high-performance teamwork, and opportunity for technical growth and extraordinary mission success is unparalleled, and as such will always be a draw for and an opportunity to develop new generations of engineers and scientists.

The many triumphs of FBC provide proof that innovative approaches can deliver mission success if the circumstances, institutional management, and project leadership require and enable it. It's very interesting to note that the essence of exploration is also "Take risk, don't fail!" All great explorations have taken enormous risks and the successful ones found creative ways to manage that risk—along with a lot of courage and some luck. Today's explorers are no different.
—B.M.

## FLASH GORDON'S TRIP TO MARS

*Following the Martian Trend in 1936*

One of the most successful movie serials of 1936 was *Flash Gordon,* based on the comic strip character created two years earlier by Alex Raymond. And since the Hollywood of the 1930s was no different from the Hollywood of today, a sequel was inevitable.

Work soon began on an all-new serial to be called *Flash Gordon's Trip to Mongo.* Mongo, ruled by the evil Emperor Ming, was a runaway planet that had careened into Earth's solar system. But in the time between the decision to make the new serial, and the day filming actually began, something unexpected and completely unprecedented happened . . .

On October 30, 1938, Orson Welles's radio presentation of *The War of the Worlds*—based on the H. G. Wells novel about a Martian invasion—electrified America. Suddenly Mars was a high-interest hot topic, and Hollywood, as always, was eager to capitalize on a trend. Thus, almost overnight, *Flash Gordon's Trip to Mongo* became *Flash Gordon's Trip to Mars.*

Hollywood never changes. Flash Gordon began his career traveling to the mysterious planet Mongo. But when Orson Welles's 1938 radio broadcast of *The War of the Worlds* brought Mars to the public's consciousness, Mongo was quickly replaced in the latest Flash Gordon movie serial by the planetary flavor of the day. COURTESY THE KOBAL COLLECTION.

There was no attempt to include any semblance of scientific accuracy, though, since that had never been the point of the serials. Consequently, onscreen Mars ended up looking remarkably like Planet Mongo, which bore more than a passing resemblance to the Universal Studios backlot.

Scientist Matt Golombek. "We each had ownership of different portions of the mission."

Matt Golombek took his ownership position seriously.

To begin with, as a scientist, he took strong exception to the whole idea of spending so much time, money, and effort just to demonstrate technology.

"We had huge fights through the mission development," he remembers, "about what constituted mission success. This is a great example of how the views of a scientist and an engineer differ. There is, actually, this formal success document that is agreed to between the project and NASA Headquarters." In an early version of this document, which sets forth the Level One requirements, each mission goal was given a percentage score that theoretically allows an overall success rate to be determined.

But in the beginning, Matt recalls with a scientist's indignation, "The success criteria that was adopted for *Pathfinder* was seventy percent if you just land safely."

Now all his passion comes to the fore as he expresses his opinion of that. "All that means is a telemetry bit from the spacecraft that says we landed safely. No picture. That's nuts! Who the hell is going to believe you landed on Mars if you don't have a damn picture?"

As we said, Matt's passion for science and the exploration of Mars is contagious, and slowly the mission-success document began to change. First, "There was ten percent added for the first picture of Mars, to prove that we landed. Then there was another ten percent to get the rover off on the surface and do something with it." He throws up his hands with a laugh. "It didn't matter what. And then there was ten percent for all the other science assignments we were going to do."

Of course, every new milestone for success that was added to the mission's objectives meant engineering changes. And scientists asking engineers for changes invariably led to . . . well, as Matt describes it, "There's always screaming. But I would say, you know, this is silly. You're not going to convince anybody we landed safely on Mars without a picture. That picture ought to be worth *fifty* percent!"

Matt's insistence on expanding the science requirements for the mission was based on practical matters as much as passion. The mission was intended to provide thirty days of surface operation on Mars. "We were going to do everything we could. We had a full ground data system set up. We were ready to operate this vehicle. We had full operational readiness tests. Everybody worked like a dog, including all the engineers, to make sure that once we got there, we weren't going to just sit around and suck our thumbs—we're going to learn something!"

Matt's familiar smile and chuckle return. "That, to me, was what was so thrilling about *Pathfinder*—that largely through my being there in the beginning, I was able to influence this project from an entry, descent, and landing demonstration piece of [insert your favorite expletive here], into a real science mission that acquired gobs of new information that has real importance to our science."

To Matt Golombek, that was the real challenge and promise of *Pathfinder*. "Here was an opportunity—an opportunity to take something that was virtually a castoff, a throwaway for the science community, and make it into something that would really stand.

"There're lots of things that went into it, not just me, but we got the science. *Pathfinder* became a science mission in every way, shape, and form."

And the reason why *Pathfinder* became a double-barreled success, achieving not only its original goal of demonstrating new technology, but also delivering a wealth of scientific information still being analyzed today, is a direct result of Matt Golombek's biggest arguments and hardest fight.

To the engineers, the goal was landing safely on Mars.

But to the scientists, the more important goal was landing on a *specific,* scientifically interesting, part of Mars.

Which brings us to the next step on our journey.

Mars itself.

# CHAPTER THREE

# "The Overarching Objective"

## Why Go at All?

*Forget rocket-ships, super-technology, moving sidewalks, and all the rubbishy hope in science fiction. No one will ever go to Mars and live. A religion has evolved from the belief that we have a future in outer space; but it is a half-baked religion... Our future is this mildly poisoned Earth and its smoky air... There will be no star wars or galactic empires and no more money to waste on the loony nationalism in space programmes.*

—PAUL THEROUX, *SAILING THROUGH CHINA*, 1983

## WHY BOTHER?

So what's the big deal about Mars?

A hundred years ago, it was understandable why people got excited about the Red Planet. Astronomers claimed to see canals, canals meant intelligent beings, intelligent beings meant neighbors. What better reason could there be to travel to another planet?

But fast forward to the *Mariner 4* flyby of 1965, and those canals disappeared forever, taking the Martians with them. Jump ahead to the Viking orbiter missions of 1976, and by then it was obvious that the vast areas of Mars that changed color on a seasonal basis weren't fields of alien vegetation, either. They were merely terrain regularly covered with dust when seasonal winds blew from one direction, and then uncovered when the winds blew from the other direction. Since the windblown dust was a different color from the surface rocks, the terrain's color changed with the seasons. End of story.

Then, in that same year, there were the Viking lander missions. The two lan-

*Visit to Utopia.* Noted aerospace artist Pat Rawlings created this painting to commemorate the fifteenth anniversary of the *Viking* landings on Mars. It depicts astronauts arriving at the *Viking 2* landing site. When humans do journey to Mars in person, the study of early landers exposed to the Martian elements for known periods of time will provide important information that could help in the development of Mars habitats. COURTESY NASA/PAT RAWLINGS.

ders, not designed to move around on the Martian surface, are still exactly where they set down and where they will remain until they go on exhibit at a colonial museum on Mars or are shipped back to Earth for display at the Smithsonian Institution (which, by the way, is the legal owner of all of NASA's "used" space equipment scattered throughout the solar system).

But those *Viking* landers had robot arms for digging into the Martian soil and special equipment to test for evidence of life. Because in 1976 there were many scientists who felt the possibility of life on Mars was a reasonable hypothesis to explore.

In fact, on the day the first *Viking* lander touched down, noted scientist Carl Sagan even held out the faint hope that its cameras might show "Martian critters" walking past on the horizon. At the very least, other scientists were hopeful that the first images sent to Earth might reveal small but hardy Martian moss or lichens.

Visually, though, *Viking* found nothing. (And for all those in the back shouting, "Face on Mars! Face on Mars!" we'll be getting to that. But as Dr. Zaius said to astronaut George Taylor at the end of *Planet of the Apes,* "You might not like what you find.")

The lack of visual indications of life from the *Viking* landers meant that, in fairly short order, the search for life on Mars had progressed from "critters" to plants to microscopic organisms—you know, relatives of the germs and bacteria which on Earth had killed the Martians in every version of *The War of the Worlds.*

Here, it becomes important to remember that each *Viking* lander carried three instruments designed to perform three different experiments not to detect life directly, but the biological processes associated with life. That distinction, which might seem like hairsplitting, is very significant to scientists. Life is such a big concept, after all, and on Earth there are so many different forms of it that looking for one type might prevent you from seeing another. But one common characteristic all life-forms share—presumably, even Martian life-forms—is a set of underlying biological processes that allow living things to live.

What kind of biological processes?

There's no simple answer, because any attempt to define those processes invariably involves everyone in trying to define an even larger question: What is life? And that debate is still going on today.

But in the early 1970s, when part of that debate centered on what instruments should be included on the *Viking* landers, here's what the exobiologists faced— exobiologists being those optimistic scientists who specialize in the study of life everywhere in the universe, except for Earth. (The term *exobiology,* a staple of science fiction, was coined in 1960 by Joshua Lederberg, winner of the 1958 Nobel Prize in Physiology or Medicine.)

Before they could even choose their instruments for their landers, the Viking mission planners needed at least some kind of definition of life.

In 1965, in a NASA publication plainly titled, *An Analysis of the Extraterrestrial Life Detection Problem,* three NASA researchers had proposed five characteristics that all living things shared: growth, movement, reproduction, metabolism, and

**THE FIRST COLOR CLOSE-UP OF MARS** Computer processing was much slower in 1965, and as the historic first close-up images of Mars were slowly being transmitted back from the *Mariner 4* probe, the engineers and technicians at JPL were too impatient to wait for the final processed images to be printed. Instead, they took strips of paper on which the numerical information for the first image was printed out and hand-colored the numbers according to the color values they represented, like a paint-by-numbers picture. The result was the first color image from Mars, which is still on display today at JPL. COURTESY NASA/JPL

ONE LAST GLIMPSE OF OLD MARS Released in 1964, one year before *Mariner 4* consigned every science-fiction film about Mars to the realm of fantasy, *Robinson Crusoe on Mars* made a valiant attempt to introduce at least a hint of scientific plausibility to its presentation of the Red Planet. Unfortunately, few movies made since have attempted to keep up with the scientific realities of Mars, leaving us with two completely different visions—Mars as it is, and Mars as it can never be. COURTESY PARAMOUNT PICTURES/THE KOBAL COLLECTION

**THE SCAR OF MARS** Valles Marineris is one of the largest geological features on Mars, and the largest canyon complex in the solar system. COURTESY NASA

**THE REALLY GRAND CANYON** This is a composite image of Valles Marineris on Mars—a canyon complex that is more than 1,800 miles long and about 3.5 miles deep. It is so large, in fact, that it is one of the few features on Mars whose position actually matches a "canal" on Percival Lowell's maps of Mars. Valles Marineris was first photographed by the *Mariner 9* orbiter, which is the source of its name—"Valleys of the Mariner." Though the Grand Canyon on Earth was carved by water, Valles Marineris is most likely an enormous fracture rift, or fault, associated with volcanic activity. Similar fracture faults on Earth include the water-filled Gulf of California and the Red Sea. COURTESY NASA

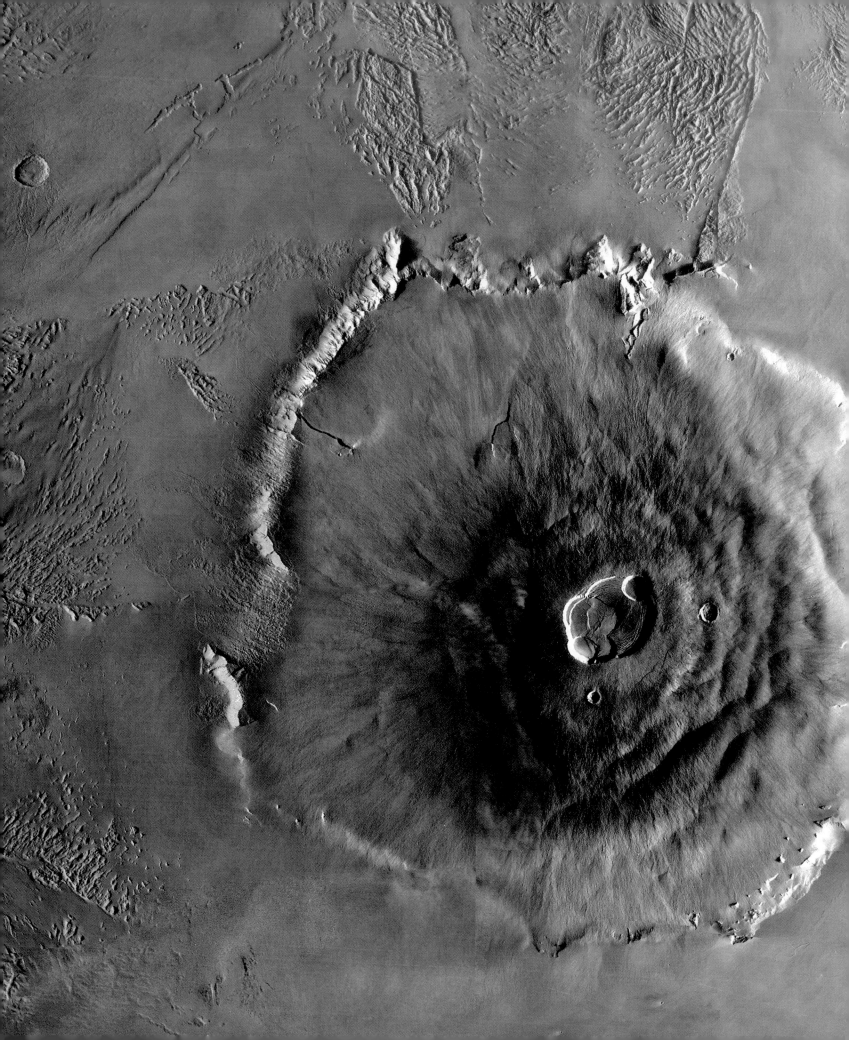

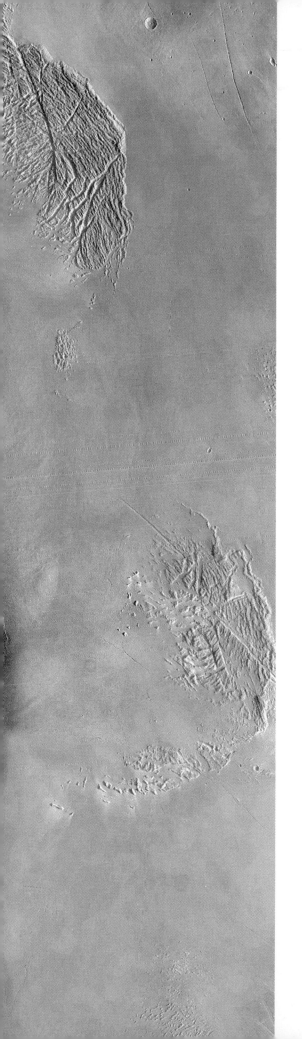

ON THE BRINK OF DISCOVERY In addition to having the largest canyon complex in the solar system, Mars also has the largest volcano—Olympus Mons, for "Mount Olympus." Early astronomers had noted a seasonal white patch of varying size at the volcano's location, which they reasonably assumed was most likely snow. Thus, the feature's original name was Nix Olympica, for "Snows of Olympus." However, when the *Mariner 9* orbiter arrived at Mars, the true nature of Nix Olympica was quickly revealed. It's a volcano with a base the size of Arizona, and so high (thirteen miles!) that clouds form at its summit—the source of the seasonal patch of white. The smaller image seen here was taken by *Mariner 9* on its final approach to Mars in November 1971. Olympus Mons is the dark spot near the top of the image. The larger image is a composite of the volcano taken from orbit. New York City would fit in its caldera with room to spare. COURTESY NASA

"I WISH I WAS WITH YOU." Astronomer Carl Sagan was a scientist and science-fiction writer who helped introduce millions of people to the adventure of space exploration through his award-winning nonfiction books for the general reader and his landmark PBS series *Cosmos*. Carl also cofounded one of the largest space advocacy organizations, The Planetary Society. In this location photo from the production of *Cosmos*, Carl stands beside a full-scale mock-up of a *Viking* lander, whose life-detection experiments he helped design. COURTESY NASA

(opposite page)
THE REAL FACE ON MARS Some conspiracy-minded folks consider this to be the product of an alien civilization, passing on to us the cryptic message "Have a nice day." COURTESY NASA/JPL/MALIN SPACE SCIENCE SYSTEMS

**THE FIRST COLOR IMAGE FROM THE SURFACE OF MARS** The *Viking* Mars landers were designed to take quick black-and-white pictures of their landing pads as soon as they touched down on Mars, just in case they began to sink beneath soft sand and their missions came to an abrupt end. However, both landers touched down safely on solid ground, and with more time available, began to return higher-resolution color images of their surroundings. This image taken by the *Viking 1* lander on July 21, 1976, is the first color picture sent from the surface of Mars. COURTESY NASA/JPL

**LAST GLIMPSE BEFORE MARS** The Mars *Pathfinder* lander is safely stowed inside the flattened gumdrop-shape aeroshell at the top of the "stack" now in position atop the Delta II rocket that will send it to Mars. Half of the Delta's aerodynamic protective shroud can be seen to the right as workers at the Patrick Air Force Station prepare to seal *Pathfinder* in place. COURTESY NASA/JPL

**PACKING FOR THE TRIP** The Mars *Pathfinder* lander undergoes final preparations in JPL's Spacecraft Assembly and Encapsulation Facility-2. The center section identified by the American flag and JPL logo contains scientific instruments and the communications equipment necessary to contact Earth. The rover *Sojourner* is in its folded-up position on one of the lander petals. Each lander petal is covered in solar-energy converters that will provide energy for the lander base station. The top surface of the rover has its own solar collectors for its independent power supply. COURTESY NASA/JPL

**ALIEN FROST** Though the atmosphere on Mars is only 1 percent as thick as Earth's, it is still able to produce familiar climatic phenomena, such as dust devil whirlwinds, clouds, and as seen here, morning frost. This image was taken by the *Viking 2* lander in May 1979, almost three years after it first touched down. Speckles in the lower half of the photo are the result of random errors in the transmitted image. COURTESY NASA/JPL

Well beyond their planned thirty-day mission, the *Sojourner* rover and the *Pathfinder* lander fell silent, mostly likely victims of the thermal stresses caused by the extreme temperature fluctuations through the cold Martian days and even colder nights. Someday, though, explorers from Earth will visit the Carl Sagan Memorial Station, site of the first robotic exploration of Mars. PAT RAWLINGS/COURTESY OF NASA AND THE ARTIST

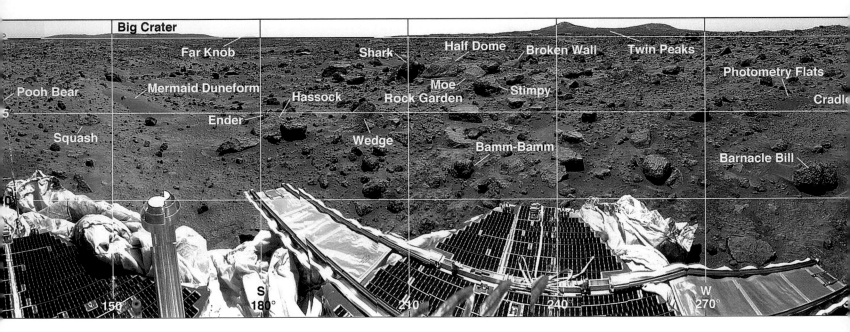

**Big Crater**

Far Knob

Shark

Half Dome

Broken Wall

Twin Peaks

Pooh Bear

Mermaid Duneform

Moe

Stimpy

Photometry Flats

Hassock

Rock Garden

Cradle

Ender

Squash

Wedge

Bamm-Bamm

Barnacle Bill

S
180°

150°

210°

240°

W
270°

THE "PRESIDENTIAL PANORAMA" This 360° view of the Mars *Pathfinder* landing site is made up of more than 300 separate images acquired over three Martian days. As geologist Matt Golombek had predicted, *Pathfinder* landed on a floodplain where an ancient deluge had washed a wide assortment of rocks from many different areas on Mars, all within easy reach. The rover *Sojourner* can be seen making its way to the rock named "Yogi." COURTESY NASA/JPL

THE NAMING OF NAMES The *Pathfinder* team chose informal and often whimsical names for the features of their landing site on Mars. But in what is rightly considered to be a classic of Mars fiction as well as one of the great short story collections of the twentieth century, celebrated writer Ray Bradbury took a more somber approach when he described how his first settlers named the landmarks of their new home in *The Martian Chronicles.* "They came to the strange blue lands and put their names upon the lands. Here was Hinkston Creek and Lustig Corners and Black River and Driscoll Forest and Peregrine Mountain and Wilder Town, all the names of people and the things that the people did. Here was the place where the Martians killed the first Earth Men, and it was Red Town and had to do with blood. And here where the second expedition was destroyed, and it was named Second Try, and each of the other places where the rocket men had set down their fiery cauldrons to burn the land, the names were left like cinders." COURTESY NASA/JPL.

**A PATH NOT TAKEN** After the success of Mars *Pathfinder*, JPL began work in earnest on a Mars Sample Return (MSR) mission. As conceived at the time, a *Sojourner*-type rover would deliver rock and soil samples to a lander based on the design of the Mars Polar Lander (MPL). A Mars Ascent Vehicle (MAV) would then carry the samples into Mars orbit, where another robotic vehicle would rendezvous with it, acquire the samples, and return to Earth. Unfortunately, in the reassessment of the Mars exploration program that followed the loss of the MPL in 1999, a sample-return mission was considered to be too complex to be affordable with existing technology, and the program was canceled. However, with the success of the Mars Exploration Rover missions of 2004, new possibilities for an MSR mission are under active consideration, with tentative plans for the first to be launched early in the next decade. COURTESY NASA

WATER WORLD These computer-generated images show what Mars might look like today if it had retained the water believed to have existed there much earlier in its history. The mystery of what happened to the water of Mars is one that is still to be solved. Some of the missing water exists as water ice, frozen deep beneath the planet's dry surface. Additional amounts lie trapped in the north and south polar caps, which also consist of frozen carbon dioxide, better known as dry ice. Unlike Earth, Mars does not have a strong magnetic field to shield it from the catastrophic effects of solar flares. Based on information returned by the Mars *Odyssey* orbiter, NASA studies of the effects of massive solar flares ejected by the Sun in October 2003 suggest that large amounts of Mars's atmosphere and the water vapor it contained might also have been blasted into space over millennia. It is crucial to understand what happened to Mars's water, in order to know if the same process might someday affect Earth. COURTESY NASA/GREG SHIRAH

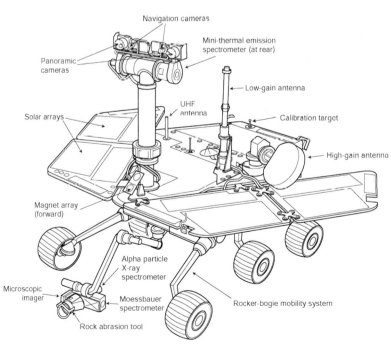

Navigation cameras

Mini-thermal emission
spectrometer (at rear)

Panoramic
cameras

Low-gain antenna

Solar arrays

UHF
antenna

Calibration target

High-gain antenna

Magnet array
(forward)

Alpha particle
X-ray
spectrometer

Microscopic
imager

Moessbauer
spectrometer

Rocker-bogie mobility system

Rock abrasion tool

**THE NEXT GENERATION A Mars Exploration Rover, 2004.** COURTESY NASA/JPL

**SOME REASSEMBLY REQUIRED** Like its smaller cousin, *Sojourner,* a Mars Exploration Rover had to make its journey to Mars in a folded-up configuration in order to fit the tight constraints of the lander. In this photo, the rover *Spirit* sits atop its lander base with its solar panels folded up and its wheels in their stowed position. The aeroshell that will protect *Spirit* during its entry into the Martian atmosphere can be seen in the background. COURTESY NASA/JPL

+X

S/N 001

MORE TESTING During the extensive testing process, not all parts of the Mars Exploration Rover (MER) hardware have to be tested at the same time. This photo shows an engineering model of a MER traveling down a deployed ramp in the Mars "sandbox," without its solar panels and robotic arm attached. COURTESY NASA/JPL

**COLUMBIA MEMORIAL STATION** Sixteen Martian days after it touched down, the rover *Spirit* looks back at its lander, showing how its ramps deployed over the retracted airbags. Unlike the Mars *Pathfinder* lander, the MER lander has no scientific instruments or communications equipment. Instead, the Mars Exploration Rover carries everything with it, allowing it to explore far afield. COURTESY NASA/JPL/CORNELL

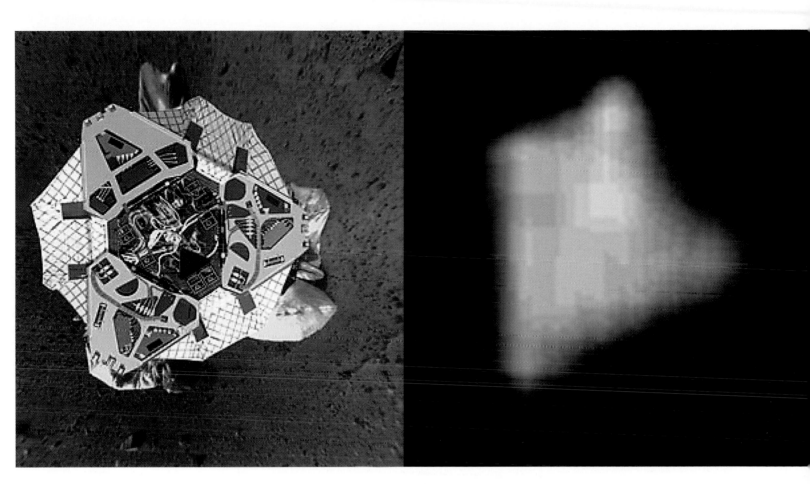

EYE SPY ON MARS To the left is a computer-generated image of the *Spirit* lander, assembled from multiple images taken by the rover. To the right is an actual image of the lander taken by the camera onboard the Mars Global Surveyor orbiter. Images returned by NASA's and JPL's next Mars orbiter, Mars *Reconnaissance,* will be able to capture more detail. COURTESY NASA/JPL/CORNELL/MALIN SPACE SCIENCE SYSTEMS

GOING LOONEY ON MARS After years of serving as an unofficial ambassador of Mars exploration, Looney Tunes character Marvin the Martian finally received official recognition when he became the mascot of the Delta launch team responsible for sending MER-A (*Spirit*) to Mars. Not to be outdone, Daffy Duck, in the guise of heroic space explorer Duck Dodgers (in the twenty-fourth-and-a-half century), became the mascot of the MER-B (*Opportunity*) Delta launch team. COURTESY WARNER BROS./NASA/BOEING/JPL/CORNELL/ACME DISINTEGRATION DEVICES

**BITS AND PIECES** When the rover *Spirit* acquired this mosaic image of the Martian landscape, all eyes were drawn to the white speck about 650 feet away. A higher-resolution image taken the next day confirmed what the camera team suspected: The white speck was actually *Spirit*'s heatshield, jettisoned while the lander was high above the Martian surface. Further confirmation of the speck's identity was obtained by images from the Mars Global Surveyor orbiter. Before *Spirit* landed, the area where the speck is located was undisturbed. After *Spirit* landed, images of the same location revealed a large gouge. COURTESY NASA/JPL/CORNELL

STRETCHING THE TRUTH When an ancient meteor slammed into Mars millions of years ago, the explosion exposed layers of sedimentary rock formed on the shore of an even more ancient body of water. When the rover *Opportunity* explored that crater, now called Endurance, the rock layers revealed the long and complex history of the chemical nature of the water. The picture to the left shows three different sediment layers, where *Opportunity* used its Rock Abrasion Tool (RAT), in an approximation of their true color. While the layers' colors don't look particularly different to the untrained eye, engineers are able to exaggerate the subtle differences among them with a technique called decorrelation. The resulting garish colors in the picture on the right help geologists identify the differences in composition among the three layers, which now are clearly quite distinct, revealing important clues as to what conditions were present when each layer was formed. COURTESY NASA/JPL/CORNELL

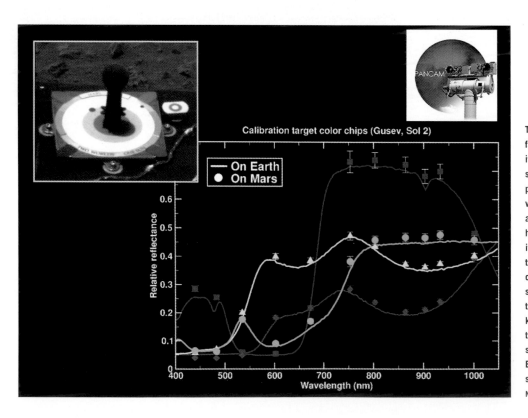

Calibration target color chips (Gusev, Sol 2)

— On Earth
● On Mars

TRUE COLORS Mars is about 50 percent farther from the Sun than is the Earth, and its thin atmosphere affects how sunlight is scattered and what kind and size of dust particles are carried aloft. Because of these visual variables, all color images from Mars are approximations, and no one will know how close those approximations are to reality until the first humans visit. In the meantime, both Mars Exploration Rovers carry color-calibration targets with known color swatches, so engineers can compare how those colors appear on Mars with how we know they appear on Earth. By ensuring that the color characteristics match, we can be sure that the color images we receive on Earth are very close to what humans will someday see on Mars. COURTESY NASA/JPL/CORNELL

**THE NEXT STEP** Following NASA's Vision for Space Exploration, the next decade and a half will be spent creating the scientific and technological foundation for a human expedition to Mars. This is one concept for the vehicle that will carry a crew to Mars, once robotic craft have established an operational and fully supplied base camp. COURTESY NASA/JOHN FRASSANITO AND ASSOCIATES/JOHNSON SPACE CENTER

SAFETY IN NUMBERS Just as in the days of the Apollo missions to the Moon, the key word for astronaut safety on Mars will be redundancy. On a mission that could last years, the first human explorers on Mars will be supplied with backup supplies and equipment, and perhaps even backup habitats. COURTESY NASA/JOHN FRASSANITO AND ASSOCIATES/JOHNSON SPACE CENTER

**THE HUMAN TOUCH** Steve Squyres, principal investigator for the Mars Exploration Rover missions, has estimated that what a rover can do in a full day, a human geologist on Mars could accomplish in thirty seconds! Rovers and robots may prepare the way, but the full exploration of Mars will not begin until humans arrive to carry it out in person.
PAT RAWLINGS/COURTESY OF NASA AND THE ARTIST

irritability. (And no, that last one had nothing to do with being cranky, it referred to a living thing's ability to respond to positive or negative conditions in its environment.)

In general, it was a good list. At least, it gave exobiologists a starting point from which to organize their thoughts. And their first thoughts identified a few problems with that list, especially in the context of sending instruments to Mars to look for those defining characteristics of all life-forms.

Movement, of course, did not apply to most plant life-forms. Over a period of years, a forest might advance or recede across a particular landscape on Earth, but an instrument on a Mars lander designed to operate for ninety days at most, wouldn't have a chance of detecting movement on such a slow scale. Also, if life on Mars moved at a different tempo than life on Earth, perhaps as a way to cope with the intense cold and lack of oxygen, those other four qualities of life would be equally hard to detect over only ninety days, as well. However, if the exobiologists put plant life to the side and turned their attention to life at a microscopic scale, then the odds of detection might go up. Since, as bacteria thrive they metabolize nutrients and produce waste products, the exobiologists reasoned that if those waste products could be detected by an instrument on Mars, then that would count as the identification of a biological process. And that would allow scientists to reasonably infer, or, to be optimistic, unambiguously prove they had detected a sign of life.

On Earth, microbes can remain in a form of suspended animation, as so-called spores, for an astonishingly long time. In May 2004, at the General Meeting of the American Society for Microbiology in New Orleans, researchers from Penn State University set a new record by announcing the discovery *and* revival of "micro-microbes" found in a core sample of Greenland ice dating back 120,000 years. Even more incredible, the researchers suggest it might be possible that the microbes had been trapped beneath the ice for *millions* of years since the ice sample that contained the extremely small microbes had been mixed with permafrost at the bottom of a glacier. Yet when those microbes were exposed to heat, water, and nutrients, they began to reproduce and form colonies.

However, some other scientists had a problem with the exobiologists' idea, and it had to do with the words "on Earth." (*Star Trek* fans know where this is leading; they've heard the phrase a thousand times: "It's life, Jim, but not as we know it.") Carl Sagan had a succinct way to describe the attitude behind such objection: "Earth chauvinism."

In other words, what could be more arrogant than expecting life on Mars to be just like life on Earth? All life on Earth is based on carbon. (In twenty-three words or less: Life requires lots of complex chemical interactions, and the carbon atom has proven incredibly versatile when it comes to combining with other atoms.)

But what if Mars life was based on another element, like silicon? Not quite as versatile as carbon, but a favorite of science-fiction writers. Could any exobiologist say what kind of waste products a silicon-based microbe would produce? Could any exobiologist say what kind of nutrients silicon-based life might require?

The answer, of course, is no. And back then, it was such a resounding *no* that

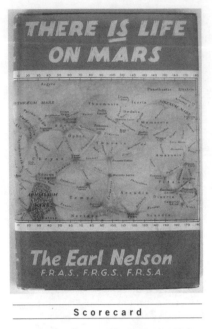

## THERE IS LIFE ON MARS

*The Common Wisdom of 1956*

In 1956, the public's interest in Mars once again reached a peak as the Red Planet and Earth made their closest approach in seventeen years. Published to coincide with that event, *There Is Life on Mars* was written by the Earl Nelson, the great-great-grandnephew of England's naval hero Admiral Horatio Nelson.

This book was a follow-up to the earl's 1953 volume *Life and the Universe* and was a summary for the general reader of all that was currently known about Mars.

At the time, subtle seasonal changes in color suggested that a type of vegetation might cover the Martian deserts. Also, the questions raised decades earlier, since 1894, by Percival Lowell's theory of intelligently constructed canals still had not been definitively answered. By the mid–twentieth century, though, most scientists doubted Lowell's proposition.

Of the possibility of life on Mars in 1956, Lord Nelson had this to say, along with a grim prediction about the fate of the first human expedition to the planet.

those questions about being able to detect life "not as we know it" were simply put to the side. After all, how can we look for something without knowing what it is? And, more basic yet, how would we ever know we had found it?

Back in the 1960s, one of the leading voices of exobiology was Wolf Vishniac, who in addition to having one of the coolest names in space exploration, was a microbiologist at the University of Rochester. He also had a knack for cutting to the chase. If exobiologists insisted on trying to imagine an infinite number of hypothetical chemistries that might give rise to alien life-forms, they'd *never* accomplish anything. So Vishniac made the blanket assumption that life "always will be based on carbon." (However, because he was a scientist, he also realized his assumption might have to be revisited one day, so he added, "It may turn out that we are deluding ourselves—that we are simply limited in our imagination.")

But Vishniac's overall assessment of how the search of life should proceed carried the day. The still-lively debate that followed, even after *Mariner 4* revealed how inhospitable Mars was in terms of temperature, atmospheric pressure, and ultraviolet radiation, was confined to detecting signs of Martian life based on the assumption that at its most basic level, it would be similar to Earth life.

The end result of more than a decade of study and passionate debate became the three science-instrument packages carried by each *Viking* lander to perform three different types of life-detection experiments on soil samples scooped up by the landers' robotic arms.

The conditions for success or failure were simple and straightforward (and, of course, described in more detail in a sidebar). But the short-form version is that if the experiments consistently returned one type of result, it would indicate that biological activity had been detected. If the experiments consistently returned a second type of result, it would indicate that no biological activity had been detected.

But proving that Murphy's Law works on Mars as well as on Earth, two of the *Viking* lander biological-detection experiments returned a *third* type of result, which no one had anticipated. And the third one returned a series of *positive* readings that by definition should indicate the presence of some type of microbial life, but in amounts so small that the lead scientists decided that there had to be another, nonbiological reason for the results, even though in 1976—as now—no one could come up with another explanation that everyone can agree with.

End result of the *Viking* landers' search for life on Mars?

Let's just say the old fight *still* continues, and call the *Viking* results "inconclusive."

## WHO CARES?

So why should anyone care if there is—or was—life on Mars?

This question connects to a vastly bigger one, that connects in turn with what people thought was reasonable back in 1904—the idea that Mars had canals, and thus a Martian civilization.

Imagine for a moment that Percival Lowell was right, that the images sent back from *Mariner 4*'s flyby in 1965 showed the artificial canals and domed cities of an advanced Martian civilization.

If that had happened, is there any doubt that NASA would have had astronauts on Mars within ten years, no matter the cost?

That's how much Earth's people want to know the answer to one of the biggest questions there is: Are we alone?

The logical connection between Martian microbes and alien civilizations really isn't that outrageous. In fact, the ongoing search for signs that either possibility might be true is a cornerstone of NASA's most fundamental goals.

To understand the importance of NASA's search for life "out there," we'll take a short sidetrip by way of Iran and meet Dr. Firouz Naderi, director of JPL's Mars Exploration Program.

First of all, clear your mind of your stereotypical image of a working engineer. Firouz looks like he just stepped out of *GQ* and would more likely be seen in a Fortune 500 boardroom than in an engineering laboratory. But as the leader of a $600-million-a-year program he needs to look the part.

At the senior management level, the jeans and polo shirts that form the basic JPL uniform are more often replaced by tailored suits—in Firouz's case, immaculately so. But though his public persona might appear to be more in tune with Washington, D.C., than Pasadena, California, there is no mistaking that he is possessed of the same passion for his job shared by the newest JPL employee fresh out of school, looking at his first pork chop plot or designing a mission-critical controls algorithm.

And also like many people at JPL, Firouz never imagined he would find his calling in the exploration of space.

"I am an 'outlyer,'" he says. "You talk to many of my colleagues here, and they go back to when they were seven years old and they were playing with paper airplanes or their father was in the air force. From the get-go, they knew that they wanted to be an astronaut and fly in space." But that was not Firouz's experience. "I learned to be an architect."

Fortunately, the nature of his early architectural training gave him a solid background in systems engineering—that is, the ability to combine many different subsystems and processes into a functioning whole. "So my engineering brought me into an organization which does that. And that path took me to this position. But I was not one of the people who at seven years old knew that I wanted to go to Mars."

However, Firouz *had* known he wanted to be an architect. "I think for me that was more tangible. I liked to design things and so architecture is something that appeals to me. In fact, I started as an architect, and then changed my mind and went into engineering."

Why he changed his mind, to this day he doesn't know. But he doesn't regret it. "It is hard for me to believe that I've been at JPL for twenty-four years. Still, if I go on a long vacation in Europe, at the end of the vacation, I don't have a sense of 'Damn, it's now back to work again.' I look forward to it. And for most of us fortu-

If there were intelligent beings on Mars, it is an interesting flight of fancy to speculate on their probable appearance. Invariably, the fiction writers picture them as small replicas of ourselves (the little Martian men) or as great, clumsy creatures of the most bizarre appearance, something like a cross between an octopus, an ant, and an elephant. Of one thing we can be fairly certain: they would not be small and active, with large and efficient brains, walking upright and possessing arms and hands. It is most unlikely that they would resemble us in anything but the most remote degree. Certainly they would be nothing like men in miniature . . .

When the first space travellers from the Earth set foot on Mars it will be interesting to know what they discover. Even though they, themselves, may never return, they may be able to relay the information back to Earth.* The probability is that they will find themselves on a strange and barren world, where the only visible life is some form of lichens covering areas that were once the beds of ancient seas, surrounded by tracts of red and yellow desert . . .

\* By shortwave radio.

Eight years after the publication of Lord Nelson's book, images returned by the *Mariner 4* space probe strongly indicated that the canals of Mars had been illusions, and that if there were vegetation on Mars, it did not exist in vast fields—at least, not in the areas imaged by *Mariner*.

Twenty years after the book's publication, the *Viking* landers conducted the first on-site tests for life on Mars, with inconclusive results that are still the subject of debate today.

But as for "the beds of ancient seas," the evidence provided by the MER missions of 2004 shows that that prediction has been proven true, at least to some extent.

## AN ALIEN OCEAN

*The Possibility of Life on Europa*

Of Jupiter's four largest moons, Europa is the smallest, not quite as big as Earth's own Moon.

In 1979, the first close-up images of Europa were obtained by the *Voyager 2* probe and revealed the moon's unusual surface—it was smoother than the others orbiting Jupiter, with many fewer craters.

From 1996 to 2003, more detailed studies by the *Galileo* probe confirmed what many astronomers suspected—Europa's surface was largely free of craters because it was made of ice. Any craters that formed were slowly erased by the ice shifting and by upwellings of water or slush from the ocean that exists deep beneath the moon's icy crust.

Scientists believe that the tidal forces Europa experiences in its orbit around Jupiter provide enough stress and strain to keep the center of the moon heated, allowing the liquid ocean to exist.

Thirty years ago, the discovery of an ocean on Europa would not have led any exobiologist to be confident life might exist there, though several scientists did suggest that possibility. At the time, though, all life on Earth was believed to depend on the Sun. Even creatures and organisms that lived in total darkness were part of a food chain that always had at its base plants that used photosynthesis to obtain energy from sunlight. So if there was life on Europa, cut off from sunlight under miles of ice, what could possibly fuel it?

On February 17, 1977, a team of geologists, geochemists, and geophysicists from Oregon State University, the Woods Hole Oceanographic Institution, and Scripps Institution of Oceanography made a discovery deep

nately here at JPL, it is that way. It really is not a chore. It's nice that they give us a paycheck and we can pay the mortgage."

The work at JPL might not be a chore, but that doesn't mean it's not hard. "You do have to love the work," Firouz says, because that's the only way to deal with the stress and pressure of it. And that stress and pressure doesn't just come from the need to meet deadlines and budgets. "You are in the public's eye. Our failures are as spectacular as our successes—your distant cousin six times removed knows that you are working on something that might have succeeded or failed. So you have to love it, and most of us do."

Firouz was born in Iran and first came to the U.S. to earn his PhD in electrical engineering, specializing in the just-emerging field of digital processing of television signals. On graduation, he returned to Iran to work for the country's national television agency, coming back to the United States three years later, after Iran's revolution.

Firouz had completed his graduate work at the University of Southern California, and several former students who had been in his class were now working at JPL. That was his first contact with the Lab, and it was all he needed to begin his new career.

At the time of the Pathfinder mission to Mars, Firouz was program manager for Origins, a NASA program studying the cosmos beyond our solar system, to answer two of science's most profound questions: Where did we come from? Are we alone?

Right now, NASA's great space-based observatories, including the Hubble Space Telescope, are methodically assembling the scientific information required to answer the first question, by eventually peering back in time to what might be the first few hundred thousand years of the universe's existence and gradually revealing the processes by which galaxies, stars, and planets form. To answer the second question—the search for life outside our solar system—breakthrough technology is being developed to create a new generation of space telescopes able to detect Earth-size planets around other stars, image them, and maybe even see their continents and oceans, and analyze their atmospheres to look for telltale signs of life, such as oxygen and methane—gases that exist at their present levels in Earth's atmosphere only because they are waste products of biological processes.

Two years after the success of *Pathfinder*, Firouz was still program manager of Origins when NASA faced what is invariably called, "the '98 failures"—the loss of the Mars Climate Orbiter and Mars Polar Lander on arrival at Mars, for reasons which, in hindsight, should have been avoidable.

Thomas Young, a retired executive of Martin Marietta and former Viking mission director, chaired the committee that investigated the failures. As Firouz explains, "When he looked at the Mars Program, he said that the management structure and communications channels were as much part of the failures as anything else."

One of NASA's responses to Young's report was to establish two managers responsible for integrating the agency's plans for exploring Mars: one manager at JPL in Pasadena, California, and one at NASA Headquarters in Washington, D.C. The JPL manager—Firouz—would report to the Headquarters manager, Dr. Orlando Figueroa, and be responsible for every aspect of the Mars Program's implementation. The Headquarters manager would be responsible for policy and funding.

The key to this new way of organizing the Mars Program, Firouz says, was that "in order to tighten the communications, the two managers became the single point of contact for each other, one at Headquarters and one at JPL. Until then, there had been too many points of contact, which had led to confusion and miscommunication."

Firouz remembers the date exactly: on April 7, 2000, he became manager of the Mars Program at JPL.

The new position was a perfect fit with his previous experience on Origins.

Just as each individual mission to Mars has to be seen within the context of NASA's overall program of Mars exploration, Firouz says, "I have to see the Mars Program within the context of NASA's space science. If you stand back so you can see the panorama, a very important, overarching objective of NASA's space science program is: Does life exist outside the confines of Earth? They are pursuing that answer on two fronts, and I've been fortunate to have brushed against both of these."

The first of those two fronts is outside the solar system, where the Origins Program is focusing its efforts, but the second is much closer to home.

As Firouz describes that second front, "Within the solar system, there are a few places where there is potential for life—either past or present—or potential for future life, meaning visitation by humans. So we are concentrating on those places where liquid water could have existed or could still exist, because we think that liquid water is a necessary—though not sufficient—condition for the emergence and sustenance of life."

"Necessary though not sufficient" is an important distinction. Water is absolutely necessary for life as we know it to exist, but just because water does exist somewhere, it doesn't mean that life *has* to exist there, only that it *might*.

To the best of our current knowledge, the two prime targets in our solar system where water exists today under conditions suitable for life are Mars and Europa, which is one of the major moons of Jupiter. There are several other secondary targets, as well.

Again, Firouz addresses the situation with precision. "This is a very early systematic effort. We are not going looking for 'critters,' you know, walking around the place. But we *are* looking for building blocks. That's why, for example, we also want to look at comets, because we think that the original building blocks of the solar system are preserved in them. Therefore, if we go to comets, we can access pristine samples of what the initial conditions might have been like. So comets become part of this story themselves, as we look for things related to this quest for life."

Another target is Titan, one of the major moons of Saturn, and the only moon in our solar system with a significant atmosphere (50 percent more dense than

beneath the Pacific Ocean that fundamentally changed our understanding of life on Earth: a community of sea creatures unknown to biologists, clustered around a hydrothermal vent that spewed superheated chemicals into the ocean.

For the first time, biologists saw a food chain that didn't depend on solar radiation. Instead, the organisms they discovered depended on a type of bacteria called chemoautotrophs that obtained energy by "eating" the chemicals coming from the vent.

Chemoautotrophs are now believed to have been the first type of cell-based life to arise on Earth, perhaps as early as 3.9 billion years ago. Photosynthesis, on the other hand, doesn't appear until 3.5 billion years ago.

Suddenly, those miles of ice creating permanent darkness for the Europan ocean weren't an impediment to life. If the same stresses and strains that heat Europa also create hydrothermal vents or their equivalent, then two of the most essential requirements for life are in place at Jupiter—energy and liquid water.

NASA's next proposed mission to Europa is JIMO—the Jupiter Icy Moon Orbiter—currently planned for launch in 2015. JIMO would be a breakthrough mission on several fronts, not the least being its proposed use of ion propulsion powered by a compact nuclear reactor.

Though the final architecture of JIMO has not been determined, the science community hopes its payload will include a Europa lander that will carry a Europa Surface Science Package to the moon's crust to begin the search for signs of life on yet another world.

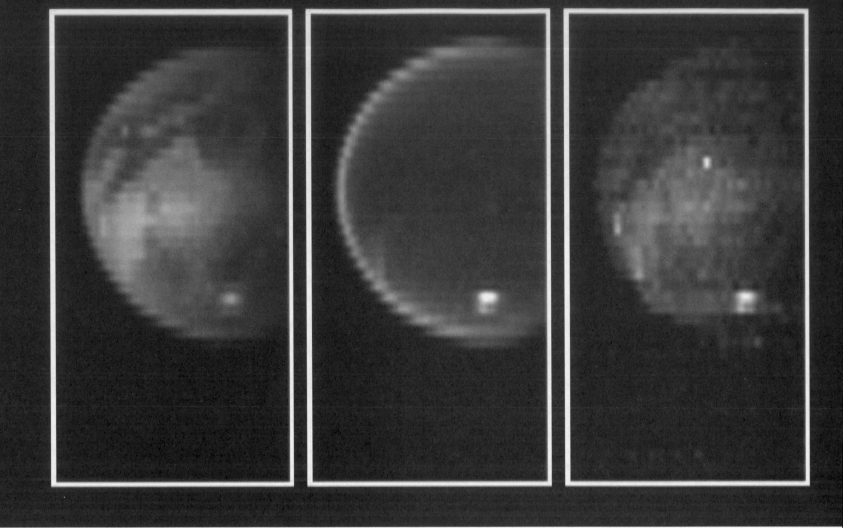

These are the first images of Titan's surface returned from the *Cassini* probe in July 2004. Because this moon of Saturn has a thick atmosphere of smog, surface detail cannot be seen in visible light, so these images were obtained by scanning Titan in the near-infrared spectrum, three times deeper than the deepest red the human eye can perceive. COURTESY NASA/JPL.

Earth's). "Again, not because we expect life there," Firouz explains, "but we believe that Titan's atmospheric conditions may be similar to what Earth might have been like in its transition from a chemical world to a biological world. We're not looking for life directly, but issues related to it."

In December 2004, at Saturn, NASA's *Cassini* orbiter will release the European Space Agency's *Huygens* probe, which twenty-two days later will enter the atmosphere of Titan and with a series of three parachutes will drift through the atmosphere to the surface. Powered only by batteries, *Huygens* will obtain striking images of a truly alien world, returning the first *in situ* measurements of Titan's unique chemistry. (*In situ* is a Latin science term meaning "in the site of," referring to collecting information at the place where the information is to be found, and not from a distance.) The probe is expected to operate on Titan's surface for less than a half hour.

But for now the most promising target in our search for life is Mars.

True, analysis of Mars by telescope, even before the *Mariner 4* flyby, showed there was no water worth mentioning in its atmosphere (though early observations, now confirmed, showed considerable water ice in its polar caps). And when *Mariner*

*4* enabled astronomers to measure the planet's low atmospheric pressure, it was obvious that liquid water could not be present on the surface of Mars. (Of course, that's another simplification because, at certain places, under certain conditions—and perhaps even recently—it might be possible for liquid water to briefly exist on Mars. Though in most places at most times, if an astronaut poured out a bottle of water on Mars, the water would instantly start to boil away into vapor and/or freeze solid, but all would disappear into Mars's atmosphere very quickly. Which is why we have sidebars.)

The new perception of Mars as a lifeless world, as suggested by *Mariner 4*, was confirmed by the next two flyby missions: *Mariner 6* arrived at Mars on July 31, 1969, and *Mariner 7*, on August 5 of the same year. (For those keeping score at home, *Mariner 5* was a successful Venus probe launched in 1967.) Recall that even though scientists were startled by the heavily cratered images they received from *Mariner 4*, it's in scientists' natures not to jump to quick conclusions (at least not publicly!), especially based on what they call "one data set," and especially when it's the first and only return of information.

But when those next two flybys of Mars showed even more heavily cratered landscapes, scientists could feel confident that their initial suspicions had been confirmed: Mars was a dead world and probably always had been.

But then we should also recall what we said about Murphy's Law working on Mars as well as it works on Earth.

It turns out that not only had *Mariner 4* just happened to fly over the most heavily cratered area of Mars, completely by random chance, so did *Mariners 6* and *7*! Even a minor difference in their arrival times at Mars—amounting to only a few hours after voyages of just over five months—would have presented an entirely different view of the planet.

In 1972, when that different view *did* become known, NASA officials said it was as if three separate probes from an alien civilization had flown past Earth, and each time had only captured images of our oceans, "proving" that our planet was completely covered by water. That's how wrong some of our initial assessments of Mars turned out to be. Not because Earth's scientists had made mistakes, but because of random chance.

This next revolution in our understanding of Mars came in January 1972 with the *Mariner 9* mission—the first probe to enter orbit of another planet. Because *Mariner 9* would have up to a year to capture images of Mars, random chance could no longer play as large a role in what this latest spacecraft might see. This was especially good news because in the last few weeks of the probe's journey to Mars, an enormous, worldwide dust storm engulfed the planet. If *Mariner 9* had been another flyby mission, its cameras would have sent back images of roiling dust clouds and nothing else.

*Mariner 9* flawlessly entered orbit on November 14, 1971, and its team of scientists and engineers at JPL were then able to wait out the storm. When the dust started to settle, the probe's mapping sequences commenced in January 1972, and scientists began to assemble detailed maps of almost the entire surface of Mars.

# WATER, WATER, ANYWHERE?

*Finding Liquid Water on Mars*

Exobiologists use an unusual term to describe the Earth's location in the solar system. They say we're in "The Goldilocks Zone." That means Earth orbits the Sun at a distance that, given the composition and density of our atmosphere, ensures our planet's average temperature is not too hot, not too cold, but just right.

Just right for what?

Liquid water.

Venus, the second planet from the Sun, lies just outside the Goldilocks Zone. Its dense carbon dioxide atmosphere of perpetual cloud keeps the planet superheated in a runaway greenhouse effect. The average temperature on Venus is about 870°F, more than 200 degrees higher than the melting point of lead. But even if Venus had escaped the greenhouse effect and had an atmosphere like Earth's, its proximity to the Sun would result in an average temperature of 212°F, exactly the boiling point of water at Earth-normal atmospheric pressure.

Mars, the fourth planet, is just outside the other extreme of the Goldilocks Zone—its average surface temperature again under Earth-normal atmospheric pressure, is estimated to be -81°F, well below the freezing point of water.

That's the trick to determining whether or not liquid water can exist on Mars. Its existence isn't dependent just on temperature, but on temperature combined with atmospheric pressure, and even with what's mixed in with the water.

The effect of atmospheric pressure on water's boiling point is easily seen in a basic cookbook. Anyone who's tried to follow a standard recipe in a high-altitude location knows that cooking times must be extended because water boils at a lower temperature. At sea level, the boiling point of water is 212°F, but in Denver, Colorado, at an altitude of about a mile, water boils at 205°F. In other words, the lower the air pressure, the lower the boiling point.

The average atmospheric pressure on Mars is less than 1 percent that of Earth's. That changes water's boiling point to about 50°F. But when scientists say "average," that implies that, at different places at different times, Martian atmospheric pressure can be higher or lower.

How high?

At the bottom of Hellas Basin, a large depression believed to have been created by a major asteroid strike, the atmospheric pressure can be double the average. In fact, about 30 percent of the surface of Mars is far enough below the equivalent of Martian "sea level" that the atmospheric pressure might permit liquid water to exist within the narrow range of temperatures between 32° to 50°F, for a short time. On a planet with an average temperature of -81°F, how likely are those relatively high readings? Actually, they're quite common. At the equator, during the Martian summer, surface temperatures as high as 80°F have been recorded.

However, though the atmospheric pressure at specific areas on Mars might remain more or less constant over time, because the thin atmosphere offers little in the way of insulation, temperatures fluctuate strongly from day to night. Early in their missions, the Mars Exploration Rovers *Spirit* and *Opportunity* experienced daily temperature swings from +1° to -150°F.

So in terms of finding the right combination of air pressure and temperature, there are many places on Mars where liquid water could exist at any given moment. However, because of the rapid temperature fluctuations, it's very unlikely that liquid water could persist for any great length of time—likely on the order of minutes to hours at most.

And that raises the question, If liquid water can exist on Mars only for short periods of time, where does it come from?

As on Earth, the answer is underground.

In some areas of Mars bordering on and including the polar regions, some geologists expect the ground will be frozen to a depth of more than a mile. But in certain areas closer to the equator, where the Martian soil receives more energy from the Sun, it's considered possible that liquid water might exist somewhere under the surface. The temperature range allowing for liquid water can even be extended to a few degrees below freezing if that water is mixed with salts (like magnesium sulfate, aka Epsom salts) that have been detected in Martian soil. Remarkably, there is evidence to suggest that even today something triggers a release of that water onto the Martian surface.

In June 2000, NASA released a series of images from the Mars Global Surveyor, showing a number of geologic features, from a few hundred to a thousand yards long, that appeared to have been formed by sudden outflows of water—each enough to fill about seven Olympic-size swimming pools by one estimate. The outflows seemed to have gushed from the sloping sides of craters, mostly in Mars's southern hemisphere.

Further study suggested that underground water, perhaps at a depth of about 1,600 feet, somehow—through heating of the surface soil, subsurface heating, or even the pent-up release of carbon-dioxide bubbles—burst out of a wall of rock, then poured down it, disturbing the soil to create distinctively alluvial-looking features.

At first, geologists roughly dated the outflows by their lack of impact craters and judged them "recent." On Mars, "recent" means any time in the past several million years. But more detailed analysis of the

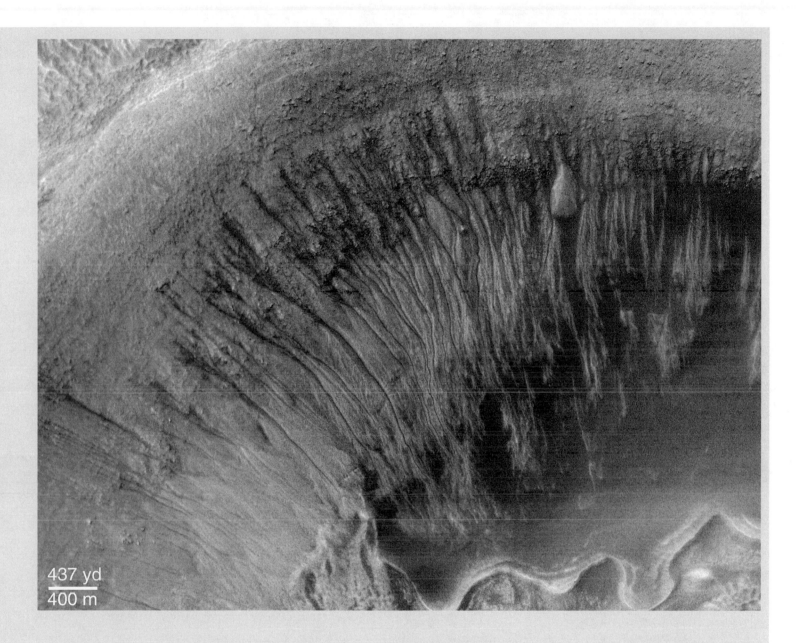

437 yd
400 m

dust layering on the material disturbed by the water raised the possibility that some of the outflows had occurred in the past few years.

Analysis of Mars by orbiters has proven that the planet holds substantial amounts of frozen water beneath its surface. The possibility that, in the present day, some of that water exists as a liquid beneath the surface as the outflows suggest, increases the chance that *if* life ever arose under more benign conditions on Mars in the past, even under today's harsh conditions Martian lifeforms might still survive in pockets of underground water.

**The crater shown here is about four and a half miles across, and attracted geologists' interest because of the many narrow gullies carved into its north wall. Detailed study suggests that the gullies were formed by flowing water carrying rock and soil debris. Each release of water indicated here is estimated to have consisted of about 700,000 gallons. COURTESY NASA/JPL/MALIN SPACE SCIENCE SYSTEMS.**

Because Mars has been geologically active in the "recent past"—which, to a geologist, means anytime in the past 100 million years or so—lava flows, volcanic eruptions, and mountain formation have effectively erased billions of years' worth of craters in certain regions, particularly in the northern hemisphere. (Polar icecaps not shown on this map.) Because the first three flyby probes had passed over the most heavily cratered areas of Mars completely by chance, the complex and varied nature of the Martian surface revealed by the *Mariner 9* orbiter was unexpected. COURTESY NASA/JPL-CALTECH.

What they saw surprised them as much as those first images of Martian craters had almost seven years earlier.

The first new features of Mars that came to the scientists' attention—because the features were so big, they literally stuck out of the clouds—were volcanoes. Enormous volcanoes, with the largest one—not only on Mars, but in the entire solar system—reaching a height of almost seventeen miles with a base that could cover the state of Arizona. (Mount Everest on Earth is about 5.5 miles high.)

The gargantuan volcano was named Olympus Mons, which is Greek for Mount Olympus, legendary home of the gods. Interestingly enough, astronomers had known for decades that there was something interesting on Mars at that location, because at certain times of the Martian year, it seemed to contain a large white patch of variable size. Giovanni Schiaparelli—the astronomer famous for noting *canali* on Mars—had speculated that the white area might be ice or snow, and had named the region Nix Olympica, for the Snow of Olympus. Detailed images from *Mariner 9*, however, finally revealed that the white area that astronomers had been seeing was actually the gigantic white clouds that formed around the volcano's peak.

But as surprising as the giant volcanoes were, no one on the science team was

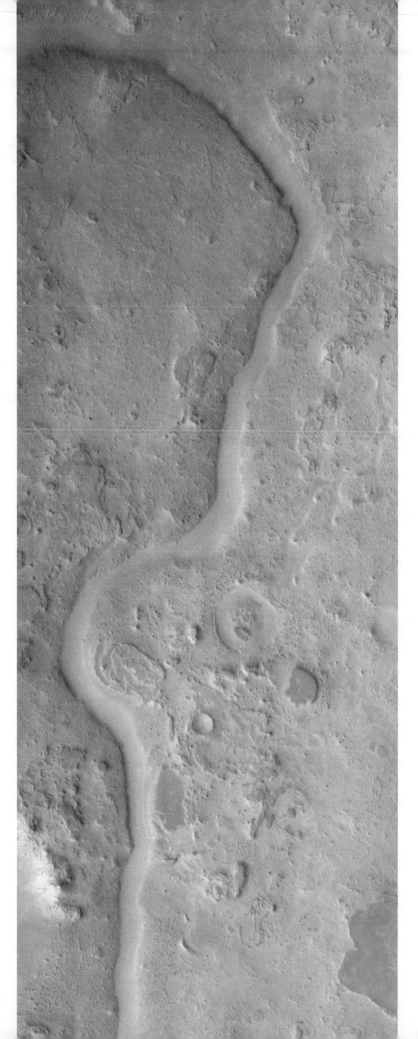

This geological feature on Mars, called the Auqakuh Vallis, is most likely a river channel carved at a time when Mars was much warmer and wetter, with a denser atmosphere. Detailed study of this image tells geologists the layers of rock the river has exposed are similar to those found in Earth's own Grand Canyon. The portion of the channel seen here ranges from about a third to a half mile in width. *Auqakuh* is the word for Mars from the Quechuan language of the Incan Empire. *Vallis* is the Latin word for valley. COURTESY NASA/JPL/MSSS.

prepared for *Mariner 9*'s next unexpected discovery on Mars: dry river valleys. Or, to be precise, geographic features that *appeared* to resemble dry river valleys.

See how cautious scientists can be?

The atmosphere and temperature on Mars made the existence of open, running water impossible (mostly). Yet some areas on the surface of Mars *appeared* to have been eroded by open, running water. Until the two contradictory observations could be reconciled, scientists had no choice but to abandon "Earth chauvinism" and consider the possibility that on a different planet there might be different erosion processes that produce features that *look* as if they were made by water, but weren't.

On the other hand, scientists also are comfortable with the wonderfully named theory of mediocrity, which essentially means that things are pretty much the same all over. Physics, chemistry, radiation, and all their interactions are presumed to behave in the most distant parts of the universe exactly as they behave on Earth under the same conditions. According to that view of things, if there appeared to be dry river valleys on Mars, then it was more than likely they were created by open, running water, just as they are created on Earth.

The concept was momentous. It meant that, despite conditions on Mars today, the dry river valleys (and eroded canyons and other features associated with water) indicated that at some time in the past, Mars had been different. Specifically, it once had an atmosphere that was thick enough and warm enough to allow liquid water to exist in tremendous quantity. And if water had once existed on Mars, then even if Mars is lifeless today, was it possible that in the past life had existed there, as well?

Which brings us back to Firouz Naderi and NASA's carefully laid plans for answering that question.

"Now, on Mars," Firouz explains, "since we cannot look for life directly, we actually look for water as a proxy for life, because we believe that that's a necessary condition. And the questions about water become: Was it there? How much of it was there? How long did it last? Where did it go? All questions related to water. Therefore, the strategy for the Mars Program has come to be known as 'Follow the water,' because we think that that's the trail that leads us to life."

The trick, of course, is that NASA is looking for something that can't exist under current conditions on Mars, at least not for more than a few minutes at the surface.

Firouz, however, keeps it simple for us. "On the surface of Mars, there is no liquid water. It's far too cold. The pressure is too low. So now we're looking for a proxy to a proxy—of water being a proxy to life. To do that, we take a look at minerals that only would have been formed in the presence of long-standing water. And that is something that we can measure today.

"So, we have a balanced strategy of looking at Mars from orbit—to give us the global context and also to point us to promising sites to go in and do *in situ* investigations. Then we land at those promising sites with instruments that are capable of detecting the type of minerals that would have been formed in the presence of water. For example, the two sites that we have gone to with MER."

*Spirit* and *Opportunity*, the two Mars Exploration Rovers that landed on Mars in January 2004, are critical links in NASA's plan to follow the water, as well as being the next generation of robotic explorers descended from the intrepid *Sojourner* rover carried to Mars on the Pathfinder mission. And though *Sojourner* was the first rover to land and operate on Mars, it in turn was the result of decades of rover development and experimentation at JPL, dating back to plans from 1963 to put a lunar rover on the Moon with a *Surveyor* lander.

But with the success of the *Viking* landings in 1976, NASA began a series of specific development programs to determine how best to design a rover that could operate autonomously on Mars. In addition to the tough engineering questions related to wheels, suspension, hazard avoidance, and power, it was important to know what kind of instruments the science community would most like to see used on Mars and then figure out how and if they could be carried by a rover.

As a technology demonstrator mission, Mars Pathfinder proved that NASA's approach to an autonomous Mars rover worked. However, because of the small size of *Sojourner*, in addition to the cameras it needed for navigation and hazard avoidance, the little rover was able to carry only a single scientific instrument.

Thus, the logical next step was to send a larger rover to Mars, one that could carry the entire suite of instruments scientists wanted to use on the planet's surface. In 1997, NASA took that step by awarding a $17-million contract to Cornell University to develop that suite of instruments, called the Athena Science Package.

The international team of Athena scientists was headed by Dr. Steven Squyres, professor of astronomy at Cornell. Seven years later, America and the world would come to recognize Steve as the high-spirited "Mars guy" who, among many other prominent public appearances during the MER missions of 2004, made the exciting and historic announcement from NASA Headquarters that the *Opportunity* rover had, indeed, fulfilled its mission to "follow the water" by discovering unambiguous evidence of long-standing liquid water, an open lake—perhaps even the shore of an ocean—on Mars.

Steve's journey from Athena to that announcement had taken seven years, but his own journey to Mars had begun even earlier, starting when he was eight years old.


## THE POWER OF PREDICTION

Steven Squyres grew up in New Jersey during the Mercury-Gemini-Apollo years, and he remembers, "As a kid, I was always fascinated by all kinds of exploration, not just space exploration, particularly. I loved reading about Arctic and Antarctic exploration, deep-sea exploration, all that kind of stuff. And I always sort of had this vision of being an explorer and going places where nobody had ever been before." He laughs as he realizes the science-fiction source of that phrase. "So by the time I got to deciding where to go to college and what to study, since I had a knack for science, I naïvely decided I'd go into geology as a way to get paid for finding new mountains. Geology was a way of combining my interest in science with my desire to be an explorer."

That "knack for science" is something that has always been with him. "I cannot remember a time when I didn't consider myself to be a scientist. I mean, when I was six years old, I didn't even have the mind-set: I'm *going* to be a scientist when I grow up. I had the mind-set: I *am* a scientist." But with more laughter, he hastens to add that at age six, "I wasn't a very good one."

However, even at that young age, Steve clearly remembers doing science experiments. "First, doing things with my father. He was a PhD chemical engineer, so he was encouraging of such things. But very soon I moved on to a lot of my own stuff. I had my own laboratory setup at home. I got my first telescope when I was eight or nine. I wasn't focused on space or geology. I was doing all sorts of things—biology, chemistry, whatever struck my fancy."

One of the young scientist's most ambitious projects introduced him to the hallmark of good scientific research—the ability to make predictions. "I spent one whole summer building this elaborate weather station for a meteorology convention, and four times a day, all summer long, I predicted the weather. Turns out predicting the weather in New Jersey in the middle of summer—that's actually pretty easy."

However, there was another area in which prediction was much more difficult, and it had to do with Steve's first telescope—a three-inch Newtonian reflector purchased from the company that has probably done as much to inspire young scientists and engineers as the magazines *Popular Mechanics* and *Popular Science*—Edmund Scientific.

Steve remembers that he began to use his telescope in January. "I went out every night for a week and looked at Jupiter through the telescope. I could see Jupiter and its moons." Like any good scientist, young Steve kept track of his observations, making drawings of what he saw, showing the positions of Jupiter's four largest moons—Callisto, Europa, Ganymede, and Io—just as the Italian astronomer Galileo Galilei had, when he discovered them with history's first astronomical telescope in 1610.

But then Steve faced the same quandary Galileo had faced. Each moon had a different orbital period around Jupiter, ranging from almost seventeen days for Callisto, to less than two days for Io. And because in a small telescope, each moon was only a point of light, as Steve puts it, "I was horribly frustrated by my inability to tell which moon was which."

Still, the allure of astronomy outweighed his frustration. "It was wonderful to go out every night and see these things move in this apparently random sort of fashion. I thought it was kind of cool."

Steve then did something else all good scientists do—he conferred with a colleague, in this case, his dad.

"My father was a chemical engineer, but he was also one of DuPont's very first computer programmers. He said, 'You know, I can probably write a computer program to help figure out which one is which.' So we sat down with a ruler, measured where everything was, and then he got a table of orbital parameters for the

moons and went off and wrote a program. This was all very mysterious to me at the time, and while he was doing that, it was cloudy for three nights in a row, so I didn't get any observations.

"Finally, he came home with this big stack of green-and-white-striped paper and said, 'Okay, I ran the program. Here's the result. I think I know which moon is which.' And we labeled them all.

"Then, here was the really cool part. He said, 'Okay, if we've got this right, this is where the moons should be tonight.'

"A prediction! A testable prediction!"

As if he's a child again, filled with that same excitement, Steve describes what happened next. "I remember going outside and setting up the telescope. I had the printout in my hand. And I looked in the telescope. And the moons were right where the computer program said they ought to be!"

In that instant, Steve experienced the true wonder and irresistible pull of science: "The power of science not only to enable you to comprehend, but to comprehend so well you can predict the future. I thought that was incredibly cool."

Of course, Steve doesn't owe his early scientific success only to his father. "My mom tolerated all kinds of nonsense. I would stink up the house with various experiments, and she'd come home and find twenty frogs in the sink." But Steve's early scientific endeavors were always encouraged and supported, setting him on the path to his future success.

By the time Steve reached college, though, that path to success wasn't quite as clear as he had hoped. "What I found was that after a few years of geology, I came to realize that despite the fact that geology as a field did offer the potential to travel to exotic destinations and see exciting things and climb mountains, I also discovered to my dismay that the geologists who've been crawling over the surface of this planet for the last few hundred years have actually done a pretty darn good job of figuring stuff out." Steve quickly adds, "That's not to say that there aren't interesting questions still out there to be addressed, but geology to me was feeling too much like filling in details. It just didn't do it for me."

Steve then applied his scientific problem-solving approach to determining his future direction. "I wanted a blank canvas to work on, and the Earth was not it. So by my junior year at Cornell, I was leaning towards something involving deep-sea exploration, deep-sea geology, marine geology—something that's considered a hard place to get to. But it still wasn't the right place for me."

So far, Steve had been unknowingly traveling along his path to Mars, propelled by his "knack" for science, his love of exploration, the encouragement of his parents, and his academic accomplishments. But his goal was not yet in sight.

That's when another aspect of science came into play, one we've seen several times in regard to the exploration of Mars.

Random chance, also known as dumb luck.

To earn his undergraduate geology degree at Cornell, Steve had to meet a language requirement. There were two ways to satisfy that requirement: take four semesters of a language or get a score of 560 on a language test.

Steve, rather unexpectedly as he recalls, achieved the required score after only three semesters of Spanish. "So what that meant was, at the last moment, I had a hole in my schedule."

The year was 1977, and that's when luck stepped in.

"I had a friend visiting Cornell at the time, and I was showing her around. We were in the Space Sciences building one day, and tacked up on the bulletin board was this little three-by-five card saying that a guy named Joe Veverka was going to be teaching a course on the results of the Viking mission to Mars. The *Viking* landers and orbiters were operating at Mars at the time, so I thought, That sounds cool."

Dr. Joseph Veverka had gone to Mars with *Mariner 9*, his first association with a Cornell-NASA team, and had returned as a member of the Viking imaging team with Carl Sagan. The course he taught was likely to be more than "cool."

Provided Steve was allowed to take it.

"My first day of class," Steve recalls, Joe asked, "Are there any undergraduates here? One hand goes up. Me. And he gives me this kind of sour look and says, 'Come see me after class.'"

Because it was a graduate course, Steve resigned himself to the fact that he was going to be tossed out. "But at the last minute, a student who knew me and who knew Joe came over and vouched for my dubious nature. So Joe let me stay in the course."

Though he didn't know it at the time, Steve's countdown to Mars had begun.

"After about three or four weeks, I decided I'd better start thinking about what I was going to do for a term paper. Because it was a graduate-level course, the paper was expected to be some kind of original—or, at least, quasi-original—research." As his first step in choosing a subject for his paper, Steve asked Joe for the key to what was called "the Mars Room."

"This was in the day before CD-ROMs, the Internet, and all that kind of stuff," Steve says, "and the Mars Room was the place where all the pictures were kept. I'd never seen pictures from a planetary mission in my life. I mean, I knew about *Mariner 9*, I knew about *Viking*, but I hadn't followed them.

"I still remember that afternoon like it was yesterday. I thought I'd sit down and just look through a few pictures for fifteen or twenty minutes and come up with a topic for my term paper. And I walked out of that room four hours later, knowing exactly what I wanted to do with the rest of my life. I mean, that was it. That was it."

While childhood memories can be difficult to access in order to recall why a particular interest was appealing, Steve has no problem remembering what was so special about that afternoon spent in the Mars Room.

"I didn't understand what I was looking at, but the beauty of it was that *nobody* understood. Nobody had seen most of the stuff before. I mean, unless you were on the *Viking* orbiter imaging team, you hadn't even looked at these pictures. So here I am, sitting on the floor, flipping through a notebook, and now I'm exploring someplace that nobody's ever been before!" Which was exactly what he had been looking for.

Steve's new focus on Mars meant changes in what he had, up till then, thought the direction of his career would be.

"By making that leap to robotic space exploration, it meant giving up the ability to physically go there myself, which is part of what had drawn me to geology in the first place. But it was a good trade to make."

Thus, for graduate school, Steve switched from geology to planetary exploration. Suddenly he found himself with a new colleague and mentor. "Carl Sagan asked me if I wanted to be his graduate student for *Voyager*, which sounded like a good deal."

"Good deal" is an understatement if ever there was one. The *Voyager* probes were to undertake the most far-reaching missions into space NASA has ever attempted, even to this day.

---

*Voyager 2* was launched on August 20, 1977, *Voyager 1* sixteen days later on September 5. Both undertook a "Grand Tour" of the outer planets, made possible by a planetary alignment that will not recur until the year 2157. *Voyager 1* passed Jupiter in March 1979 and Saturn in November 1980. *Voyager 2* flew past Jupiter in July 1979 and Saturn in August 1981. That probe used Saturn's gravity to change course and reached Uranus in January 1986, then Neptune in August 1989.

The trajectories followed by both spacecraft, traveling at about one million miles each day, will take them far beyond the influence of our Sun's gravity, meaning they will be the first human-made machines to fly through interstellar space. Amazingly, as of 2004, twenty-seven years after their launch, both spacecraft continue to function, sending back information about the characteristics of space more than 10 billion miles from Earth. (Where the one-way light time for communicating with Mars at the time of the MER landings was about ten minutes, for *Voyager 1*, the delay is now more than twenty-five hours, and for *Voyager 2*, more than twenty.)

Still, NASA confidently expects both spacecraft to continue to operate until at least 2020, by which time their thruster fuel will be exhausted and their radio signal will have diminished to the point at which it will be too weak to receive on Earth. Then, as far as NASA current navigation estimates can determine, in about 40,000 years, *Voyager 1* will come within about 1.6 light-years of a star called AC+79 3888 in the constellation of Camelopardalis, and in about 296,000 years, *Voyager 2* will come within 4.3 light-years of Sirius, the brightest star in the night sky.

---

So what does all this have to do with Mars?

Steve Squyres asked himself the same question back in 1978, and didn't have an answer then, either. "I was a twenty-two-year-old kid looking for a good opportunity, and Carl Sagan says he wants me to be his grad student for *Voyager.*" Steve laughs again as he recalls his career goals at the time. "I didn't need a plan! The door opened, and I leapt through!"

Because of his work on *Voyager*, though, one part of Steve's earlier experiences with science came full circle. When the probes flew past Jupiter, Steve at last got to see that planet's moons not as tiny points of light, but in full-color images more detailed than any ever seen before. And yes, the *Voyager* spacecraft found those moons exactly where they were predicted to be, just as Steve and his dad had found them, on a cold January night back in New Jersey.

## A PIECE OF P.I.

Even during his journeys among the outer planets, Steve Squyres remained interested in Mars. After his stint on *Voyager* ended in 1981, and after earning his doctorate in astronomy from Cornell in 1982, he joined NASA Ames Research Center in Northern California.

"This was the time Mars Observer was starting to look like the next big thing."

In keeping with the science community's approach to instrumentation at the time, the Mars Observer had been loaded with extras in an attempt to derive as

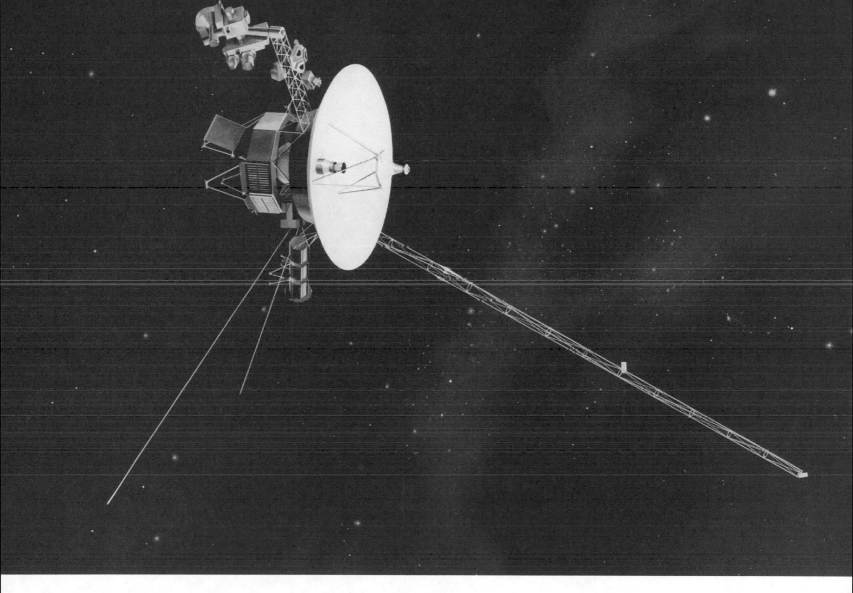

much science from the mission as possible—given the mission's high cost, it could be seventeen more years before another was attempted.

One piece of equipment in particular caught Steve's interest—a gamma ray spectrometer. The appeal of the instrument to the science community was that not only could it be used to identify the elemental make-up of the Martian surface, it could also be used for astronomical observations of mysterious gamma-ray bursts— a phenomenon that confounded astronomers until 2003.

A *Voyager* spacecraft. Today, almost thirty years after their launches in 1977, both *Voyager* spacecraft are the most distant, human-made objects from the Earth and continue to return scientific information about the nature of space, far beyond the orbit of the farthest planet in our solar system. COURTESY NASA.

A far cry from the points of light in the three-inch telescope of Steve Squyres's childhood, this montage of images shows how Jupiter and its four major moons appeared to the *Voyager 1* probe. Jupiter is the large, banded sphere. In this montage (which is not to scale), the four moons are, from largest to smallest, Callisto, Ganymede, Europa, and Io. In terms of absolute scale, Ganymede and Callisto are much larger than Earth's Moon; Io is a bit larger; and Europa is a bit smaller. The moons of Jupiter are also called the Galilean satellites after their discoverer, Galileo. COURTESY NASA/JPL.

Steve didn't know a thing about gamma ray spectroscopy, but he sensed another door might be opening for him because few others were specialists in the field when it came to using the instrument for planetary studies. "So I went off and I got myself a nuclear physics book and I worked eight months on a ten-page proposal. In the end, I got on the team."

Being on the team for a scientific instrument is an important part of coming up through the science ranks in the exploration of space. The leader of an instrument or science investigation team is called the P.I.—principal investigator. For the Mars Observer gamma ray spectrometer, Steve remembers, "At that point, I didn't feel prepared or knowledgeable enough to actually propose an instrument. I didn't have the hardware experience. I was twenty-six years old and I just didn't feel I was ready to be a P.I. on something."

The booster carrying the orbiter launched from Cape Canaveral on September 25, 1992. But after almost eleven months of cruise, when the spacecraft was about to enter Mars orbit, all contact was lost. (And to continue to keep all those stories about Mars being a "death planet" where they belong, this was the *first* loss of an American space mission in *twenty-seven* years.) Fortunately, the gamma ray spectrometer had carried out a few astronomical observations during the voyage to Mars, and so was able to return some information to its team about a gamma ray burst that just happened to occur during that time.

When a spacecraft is lost, the media often go to town with stories about the terrible waste of taxpayer dollars, as if NASA had done little more than dump several hundred million dollars into a bonfire, getting nothing back in return.

But, as with all space missions, that money was spent on Earth, mostly in the United States, over many years of research and development to come up with new instrument designs and technologies that were not lost, and served to fund the payrolls that kept a strong engineering and scientific community thriving to enable future missions. Though the spacecraft never met its scientific objectives, likely because of a flaw in the temperature-control design of its propulsion system, the knowledge gained by studying the reasons for its loss contributed to the successful and robust Mars program NASA operates today.

No one wanted to see the failure of a single mission shut down all space science for years at a time. Therefore, the stage was set for Dan Goldin's "faster, better, cheaper" era of space exploration that would tie in with Wes Huntress's ideas for the multiple low-cost missions of his Discovery program.

Also, whenever a spacecraft is manufactured, NASA often builds or orders at least one additional copy of each component. If something unforeseen happens that results in last-minute damage, a duplicate "flight-spare" is generally available for quick replacement. Because of this practical, though somewhat expensive, policy, though the Mars Observer was lost, a duplicate of its magnificent camera system was safely stored on Earth and ready for flight. Which is exactly what happened next.

In 1996, like a phoenix rising from the flames, a new low-cost mission launched: the all-new Mars Global Surveyor (MGS) spacecraft, outfitted with flight-spare equipment from the Mars Observer. In the years since achieving orbit of Mars, MGS has produced a torrent of startling and fascinating high-resolution images of the surface of Mars, becoming one of NASA's most productive science missions ever, to Mars or anywhere else.

On a more personal level, another important result of the Mars Observer mission is that Steve Squyres had served as a member of a space-exploration science team, giving him the experience he needed to finally lead a team of his own.

Jump ahead to the first press conference held at JPL on the night of January 3, 2004, after the successful landing of the Mars Exploration Rover *Spirit*, and Steve takes the stage as the MER P.I.—principal investigator for the Mars Exploration Rover missions.

Once just a kid peering at the stars through a three-inch telescope, and later a young graduate student sitting on the floor in the Mars Room, transfixed by images neither he nor anyone else could explain, Steve Squyres was now on Mars.

Which means, now it's his turn to pick up the story begun by Firouz Naderi and tell us why going there is so important.


## THE VIEW FROM THE GROUND

Steve believes there are two main reasons why the public should care about going to Mars. For someone who has always been fascinated by going where no one has gone before, it should come as no surprise that he says the inherent appeal of exploration is one of those reasons.

"I always had the hope for MER that it would play a significant role in rekindling the public interest in, or a public passion for, exploration. In *real* exploration. A lot of stuff that goes on today in the world that is called exploration really isn't. You know, somebody who skis across Antarctica backwards, that's not an explorer, that's an adventurer. I have immense respect for people who can do that, but that's adventuring, that's not exploring."

Based on his own experience, though, Steve admits that there's a reason there's more adventuring than exploring going on today. "It's the same problem I had when I was eighteen and trying to figure out what to do with my life—a lot of the good places have already been explored on Earth." Today, the situation is even more restricted. "I remember picking up an issue of *Nature* [a prominent science journal] a few months ago, and reading an article about someone who had finally gotten a submersible down to the bottom of the Arctic Ocean. And I go, 'Damn—there's another place gone.'"

But when it comes to Mars, Steve says, "I felt that with the ability to land someplace, and look off in the distance and say, 'There're the Columbia Hills! Let's go!' That this could rekindle a little public passion for real exploration." Steve believes the MER missions have made some headway in that regard. "So that's one of the reasons people should care. Because I think there is an inherent human desire to explore, and this is a way of satisfying it."

Steve's second reason for why it's important to go to Mars is directly related to what our explorations there might reveal: "Because it so directly addresses the issue of whether or not Mars was a place where life could have once arisen. And I think there are all sorts of good reasons why people should care about that."

Remember we said the logical connection between microbes on Mars and alien civilizations really wasn't that outrageous? Steve Squyres explains why.

"First of all, there's the issue of how prevalent life is throughout the universe. Right now, we've got one example: us. On this planet. And with the statistics of one, we've got no basis whatsoever for knowing how common or how rare life is throughout the universe. We just don't know. Which is the problem.

"But we do know there're lots of solar systems out there. We know there are lots of planets around stars. And there are *lots* of stars. So if we then find that life arose independently on another world, just in this one solar system, all of a sudden it takes no significant leap of imagination, or of reason or even of faith, to begin to accept the notion that life may be a common phenomenon throughout the universe."

"The other thing," Steve continues, "is one people don't think about as much, and it has to do with the issue of how life comes to be—genesis. How does that process—the creation of life from nonliving matter—how does that occur?

"That's a deep question. It happened—because here we are. But how did it happen?"

Surprisingly, if there is—or even *was*—life on Mars, then it might be easier to discover how it arose there than to discover how it arose here.

Steve explains. "The problem is, on Earth, which has been a geologically active planet for its entire history, the geological evidence of the origin of life—rocks that you can look at that tell the story—is gone.

"You look in the very oldest rocks you can find on Earth, and there's evidence of complex life. Three point eight, three point nine billion years ago, and life's there."

That evidence, arrived at after decades of careful study by geologists and biologists around the world, is no longer in question. "I feel confident in saying that by the time the Earth's geologic record emerges, life had already taken hold on this planet. Which means that the record of the *birth* of life is gone. It's just been wiped out, because there's just too much stuff going on."

By "stuff," Steve means geological activity, especially the mechanism of plate tectonics by which the Earth's crust is formed by upwellings from its molten interior, and then is drawn back down to be remelted hundreds of millions, if not billions of years later, obliterating all the fossils that might have formed in that time.

But Mars is different, as Steve points out with passion. "The beauty of Mars is that half the planet is covered with rocks that are *more* than four billion years old. So if—and this is a big 'if'—life took hold on Mars, then the record of that event may still be there to be read in the rocks. Because there are rocks that are old enough to date back to that time.

"So that's why I think people should care," Steve concludes. "Now, it's not going to fill in the potholes, it's not going to put a roof over somebody's head, and if you're not on the same page as I am—that these fundamental questions are things that people should care about—then that's probably the end of our conversation. But if you care about things like how life comes to be and whether or not it's found throughout the universe, then I think you should care about this mission and the missions that follow."

Are we alone?

Taking the first steps to discover the answer to that question is not the only reason for going to Mars, but it is one of the big ones.

And if it turns out that life once did arise on Mars, then that discovery could pose a brand-new, perhaps even bigger question.

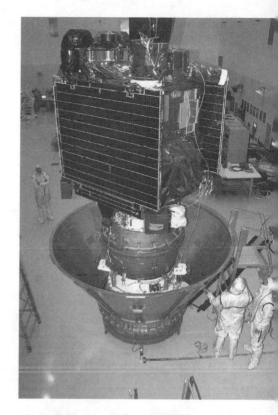

JPL technicians place the Mars Global Surveyor spacecraft in a protective canister to prepare it for transfer to the launchpad. The actual orbiter, partially hidden behind its folded solar panels, is sitting on top of the upper-stage booster that serves as the third stage of the Delta II rocket that will send the orbiter to Mars. COURTESY NASA/JPL.

These structures in Hamelin Bay, Australia, are "stromatolites"—fossilized colonies of ancient bacteria believed to be between 3.5 billion to 3.8 billion years old. Because the bacteria that formed these colonies are already complex life-forms, biologists believe the earliest forms of life on Earth appeared even earlier. Unfortunately, because of the constant recycling of Earth's crust as its surface features are melted and new crust forms, older rocks are rarely found. Though the Earth is about 4.5 billion years old, the most ancient rocks that have been discovered are thought to be just over 4 billion years old. Because these rocks were formed from molten rock and not fossil-bearing sediment, no conclusive signs of life have been detected in them. If life ever developed on Mars, which appears to have been far less geologically active than Earth, the earliest forms of that life might still be preserved in that planet's most ancient rocks, opening a window into similar processes that might have occurred on Earth. COURTESY LN LODWICK.

As scientists understand the formation of our solar system, it appears that Mars cooled and became suitable for life at least half a billion years before Earth reached the same stage. And since it is proven that meteor impacts on Mars can send Martian rocks flying into space to eventually land on Earth, if life appeared on Mars hundreds of millions of years before it appeared on Earth, it's entirely conceivable that life arose on Earth because our planet was seeded by life arriving on those first Martian meteorites.

In other words, *we* might be Martians, and by going to Mars, we're simply going home.

## THE WAR OF THE WORLDS

*The Movie*

More than fifty years after its first publication as a novel, H. G. Wells's story of an invasion from Mars still resonated with Americans living in the middle of the twentieth century.

Just as for the 1938 radio play version, the details were once again updated for this 1953 film adaptation: The action was moved from England to Los Angeles, and the three-legged Martian war machines became much-less-expensive-to-film, hovering manta ray–shape UFOs. But the overall story remained the same. Martians came to Earth, blasted everything in their path, and then succumbed to "the humblest things that God, in His wisdom, had put upon this Earth."

Since the action in *War of the Worlds* took place on Earth, depictions of Mars, accurate or otherwise, were not required.

A Martian was seen, but unlike the novel's bear-sized, saliva-dripping, two-eyed horrors, this movie Martian was a scrawny, long-limbed being with "trinocular" vision, and no logic was invoked in the film to describe why the Martians had evolved in this way. Given that many scientists of the 1950s had by then ruled out the possibility of any form of higher life existing on Mars, the idea that the aliens were from that planet wasn't at all critical to the story. It was a tale of alien invasion first, and of Mars a distant second.

In fact, by 1953, the term "man from Mars" had entered the vernacular as a term for aliens of any kind. And with interest in flying saucer and UFO reports on the upswing, there seemed to be no shortage of extraterrestrial visitors in both stories and purported sightings who were deserving of

the name, though tellingly many of them claimed to come from worlds other than Mars.

In the ongoing search for new frontiers for popular entertainment, the more we learned about conditions on the fourth planet, the more storytellers—and UFO contactees—had to set their sights on more distant worlds to maintain the illusion that anything was possible.

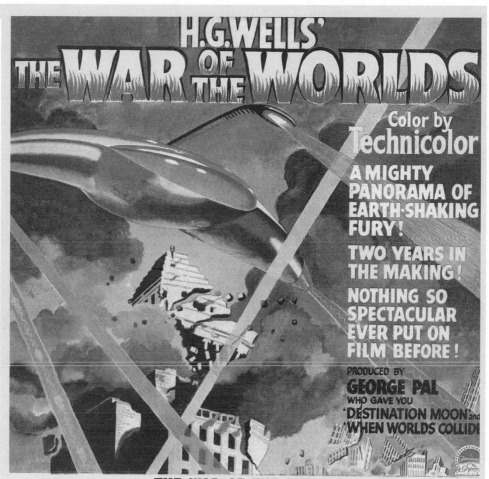

Paramount Presents **THE WAR OF THE WORLDS** Color by TECHNICOLOR
Produced by GEORGE PAL · Directed by BYRON HASKIN · Screenplay by BARRÉ LYNDON · Based on the Novel by H. G. Wells
a Paramount Picture

When H. G. Wells wrote *The War of the Worlds* in 1897, the then-common belief that the presence of canals on Mars might be signs of intelligent life couldn't be completely ruled out. By the time the movie version was released in 1953, the odds against intelligent life on Mars were fairly overwhelming. But a good story is a good story, and once again Mars served its purpose as a touchstone for the mysteries of whatever might be waiting—or in this case, lurking—in space. COURTESY THE KOBAL COLLECTION.

# CHAPTER FOUR

# "All the Things That Can Go Wrong"

## Martian Setbacks

*Mars rules catastrophes and war. It is master of the daylight hours of Tuesday and the hours of darkness on Friday. Its element is the fire. Its metal is iron. Its gems are jasper and hematite. Its qualities are warm and dry. It rules the color red, the liver, the blood vessels, the kidneys, and the gallbladder, as well as the left ear.*

—ASTROLOGICAL RULES CONCERNING MARS FROM A FIFTEENTH CENTURY MANUSCRIPT

## MERIT AND ABILITY

Sometimes things don't work out the way anyone planned, especially when it comes to Mars.

In 1954, for example, an atomic-powered Martian spacecraft sent to Earth on a scouting mission prior to a major invasion, got turned around by the London fog and landed near a pub in the Scottish highlands.

The spacecraft's crew consisted of one Martian and one giant killer robot—truly. The entire sorry affair is preserved on film as *Devil Girl from Mars*, an undeniably awful but oddly interesting British science-fiction film based on a stage play.

The premise of the story is a switch on what was for many years (and continues to be) an embarrassing standard plot for low-budget science-fiction movies: grotesque aliens arrive on Earth eager to steal beautiful Earth women, species and DNA compatibility be damned. But in *Devil Girl from Mars*, it was a beautiful Martian woman who came to Earth to steal ordinary Earth men. (Apparently, there had been a literal battle of the sexes on Mars, women won, and men were in short

supply. Sometimes, science-fiction writers inspire a generation with their dreams, and other times, not.)

For some strange reason never adequately explained in the film, the ordinary Earth men didn't want to go back to Mars to be used for "breeding purposes," and because of the gallant sacrifice of one Earth man, the Martian spacecraft blew up when it resumed its nefarious trip to London.

The film had nothing to do with Mars, except to once again use the Red Planet as a familiar touchstone for the outlandish, an excuse for the presentation of an alien perspective on our own culture. Like most science fiction of its time, *Devil Girl from Mars* is far more a reflection of a 1950s view of politics and the era's inequality of the sexes than a thoughtful projection of present or future possibilities.

That inequality of the sexes existed is not in question. In the '50s, in science-fiction films and on the covers of lurid science-fiction magazines, women were either prizes to be won or the plucky daughters of important male scientists. To be fair, from time to time, some writers did try to envision a slightly different, more inclusive future. A 1953 film, *Project Moon Base* (which was actually cut together from episodes of an unsold television series cowritten by science-fiction great Robert Heinlein), did include a female president of the United States in the far-off future of the 1970s. (Unfortunately, it also includes a female astronaut named Colonel Breiteis who, when she accidentally becomes the first woman on the Moon, must be married to the first man on the Moon by "space radio" so they can wait out the long months for a rescue flight without impropriety.)

For the most part, the roles women played in science fiction were merely the same roles women were supposed to take in society at large—even in the real exploration of space.

For example, in 1959 NASA announced its first astronauts—the *Mercury Seven*—and all were male. As a follow-up in 1960, NASA tested the country's top twenty-five female pilots, and thirteen of them passed the exact same rigorous tests required of the men. But apparently back then, being good enough to be an astronaut wasn't exactly good enough. NASA subsequently decided all astronauts should also have fighter-jet training, and since women weren't permitted to take part in fighter-jet training at the time, none of the thirteen qualified women was allowed to become an astronaut. America's first female astronaut, physicist Sally Ride, would not make her historic flight until 1983. The first female American astronaut to go into space as a pilot and as a mission commander would be United States Air Force Lieutenant-Colonel Eileen Marie Collins, in 1999. (The first woman in space was Russian cosmonaut Valentina Tereshkova, who orbited Earth in 1963. She was selected, in part, for the skills she had developed while pursuing her hobby of parachuting. In 1963, Russian cosmonauts parachuted from their capsules before the capsules reached the ground.)

Fortunately, times have changed, and no longer is half the American population

(BELOW LEFT) Astronaut Sally Ride, America's first woman in space, aboard the Space Shuttle *Challenger* on her first mission. Her favorite space movie isn't science fiction or fantasy; it's *Apollo 13.* COURTESY NASA.

(BELOW RIGHT) Lieutenant-Colonel Eileen Marie Collins, the first woman to command a shuttle mission, at the commander's station of the Space Shuttle *Columbia.* Unlike the qualified female aviators who were prevented from entering astronaut training at the beginning of the space program, Collins grew up in a different era, watching *Star Trek* and *Lost in Space,* and graduated from the United States Air Force test pilot school in 1990. A year later, she was selected for astronaut training. COURTESY NASA.

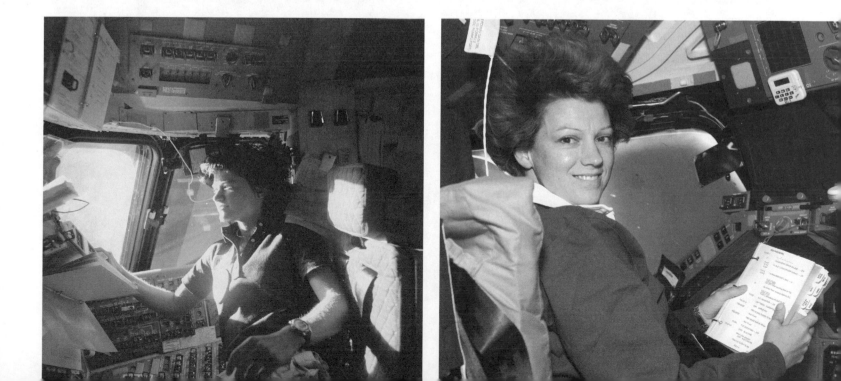

# LISTENING TO THE VOYAGERS

*The Deep Space Tracking Network*

In order to maintain communication with our vehicles in deep space, at any time of the day or night, we rely on three facilities of antennas spaced more or less equally around the globe.

These facilities and the people that design and operate them are called the Deep Space Network, otherwise known as the DSN. Each of these facilities is now made up of a set of antennas of various sizes including a huge 230-foot dish.

The first station built was an 85-foot diameter antenna, in Goldstone, California, in 1958, to support the *Pioneer 3* and *4* missions to the Moon. Its effective data rate was one thousandth of a bit per second. The second DSN station, near Madrid, Spain, and the third, near Canberra, Australia, were operational by 1967. In 1964, the first large-aperture, 210-foot antenna was installed at Goldstone to support the historic *Mariner 4* flyby of Mars. By 1988, there was a large antenna at each major facility and all the 210-foot antennas had been upgraded to 230-foot diameters to support the *Voyager* missions then traveling beyond Saturn.

These stations are linked by land communication lines back to JPL and now support data rates in millions of bits per second. The DSN also supported the Apollo missions to the Moon. Each deep-space mission built in the United States is operated through the DSN. The DSN also supports, by special agreements, international missions including those from Europe, Russia, and Japan.

One of the most exciting experiences for me at mission control during *Pathfinder* was the very first indication, shortly after launch, that the DSN had "acquired" the signal from the spacecraft. This told us that we had a working machine and that it was where we thought it should be. Every day that we were in surface operations, I stood looking at the monitor that showed us the status of what the operators at the DSN station were doing, and my heart raced when the last little red box flashed "IN LOCK," indicating we were in communication with our progeny on Mars. I never tired of seeing that marvelous process work. Here was an antenna, almost the size of a football field, rotating through the sky to point at the tiny dot of light called Mars. Behind the antenna were rooms of electronics (and people) working to sense, lock on to, and amplify a radio signal with the strength of about a trillionth of a milliwatt—the equivalent of

being able to see a candle burning in New York from Los Angeles.

The long distance cell phone call (actually, more like text messaging) record is held by *Voyager 1* (launched in September 1977). The DSN continues to communicate with *Voyager* as it approaches the edge of our solar system at a range of over 8.4 billion miles—more than ninety times the distance of the Earth from the Sun. The time to send a message and receive a reply takes more than 24 hours.

Time on the DSN stations is a precious commodity and is very carefully managed. With more than fifteen missions flying at present, the Resource Allocation Review Board (RARB) meets twice a year to negotiate with each project for the time it will get during any particular mission phase—with more detailed Resource Allocation Planning meetings held weekly. At critical times, such as a launch, deep-space trajectory changes, and landings, *Pathfinder* and MER received 100 percent coverage for a number of days, including coverage from the big 230-foot antenna. But when the landers are on the surface of Mars, the DSN typically only needs to allocate one or two eight-hour passes each day for each mission.

—B.M.

routinely excluded from most professions for lack of a Y chromosome. While it is true that science and engineering remain male-dominated fields in the United States, change is occurring. According to the National Science Foundation, in 1966, women earned only 8 percent of the science and engineering doctoral degrees given in the United States, but by 2001, that percentage had increased to 37 percent.

There are many reasons for the slow but steady growth in the numbers of women taking part in science and engineering. Certainly, a major component of change is that each new generation of women entering those fields finds more women ahead of them, working as instructors, employers, role models, and mentors. But another factor is exemplified by the policy of organizations like the Jet Propulsion Lab, which hire and promote workers not by the old-boy traditions of the past, but by merit and ability.

And that is why the world remembers one person in particular from the coverage of the successful landings of 1997's Mars *Pathfinder* and 2004's Mars Exploration Rover *Spirit*—Jennifer Harris Trosper.

## THE CIA'S LOSS = JPL'S GAIN

For *Pathfinder*, Jennifer led the development of the Attitude and Information Management (AIM) subsystem operations software, and tools, was a member of the surface operations test team, and was flight director for the spacecraft's first day on Mars. For MER, she was the project systems engineer focused primarily on risk management and the surface phase operations, then after landing became the surface operations manager for the *Spirit* rover. And in all those roles, she was a critical

Jennifer Harris Trosper cheering with Pete Theisinger in the MER support room as the first images come back from the *Spirit* rover on Mars.

member of both mission teams for the same reason everyone else was—she's a talented engineer with the drive, experience, and commitment to do what's never been done, especially when it comes to finding new and better ways to go to Mars.

Another characteristic Jennifer shares with many of her coworkers is that she didn't set out for a career in space exploration. Though, in retrospect, a favorite family story she loved to hear her father tell provides a strong clue as to where her many interests would eventually lead her.

"I think some people were engineers from age two, and I wasn't really that way," Jennifer recalls. "I grew up on a farm and played a lot of sports and really enjoyed music, and I thought I was going to be a concert pianist. That's what I wanted to be."

Indeed, Jennifer's interest in music was so keen, she started music lessons in first grade. But by the time she entered high school, she also reached an important decision point, thanks to a perceptive instructor. "I went to music camp for a couple of months and I had an instructor who was the first person who ever told me I *couldn't* do something I thought I was going to do! She said, 'You aren't good enough.'"

Jennifer had always been supported by her family in all the choices she made, and at first she treated her instructor's blunt assessment of her goal as a challenge. "I knew I had to be really good and you have to devote your whole life to being a concert pianist, so I spent a summer devoting five hours a day to practice and study." But Jennifer also had many other interests, especially sports, and by summer's end realized that she didn't want to concentrate on music to the exclusion of everything else she enjoyed.

Plus, there was that story of her father's.

"Although my father worked in the auto industry and farmed when we were in Ohio, in the late fifties and sixties he worked on the Nike Zeus and the Thor programs, so he was part of the race to the moon." (Nike Zeus was one of the first antiballistic missiles, and a forerunner of the Patriot missiles used as "theater weapons" today. Thor missiles were originally developed as offensive weapons during the Cold War, but after being retired from military service in 1964, they became an important part of the country's space effort, serving as boosters for weather, communications, and research satellites.)

Jennifer's father was a chemical engineer who loved the space program as much as he loved telling stories to his family. "I still remember a story he told me about being a test conductor on the Thor program—and he would tell it, and he would tell it again, and tell it again!"

The first operational Thor missiles were based in England, and before Jennifer was born, her parents spent a year there as her father helped with the missiles' installation. One test, however, had all the elements of a catastrophe that the god of war would have wanted. While Jennifer's father and other personnel were working outside, personnel in a Thor control blockhouse launched a missile, "and this thing goes just shooting across the *ground* and everyone's taking cover and as many times as my father's told me, I can't remember all the details, but I do remember it was very exciting. I just thought, Wow, that's so cool. That story put that kind of job in my head. Then he had this one picture in our den of the Nike Zeus missile. It was just this picture of a rocket that was hanging above our stereo and I always remember looking at that thinking, That's pretty neat."

Still, though the groundwork had been put in place by her father's stories and both her parents' encouragement, Jennifer's next career goal after concert pianist wasn't in engineering. "I was good at math and science, so when I applied to colleges and got into MIT, I figured, well, I'd better go. So I went to MIT and wanted to be an astrophysicist." Asked why she chose that new goal, Jennifer replies with the same enthusiasm with which she approaches every new challenge in her life: "Because I thought black holes were really neat! I did my high school term papers on them, new dimensions, and things like that."

An early 1960s photo of a Nike Zeus missile hanging above her parents' stereo was an early inspiration for Jennifer Harris Trosper.
© BOEING COMPANY.

But with a laugh, Jennifer recalls that her first exposure to Physics 1 at MIT led her to the realization that astrophysics was not for her (not an uncommon experience for would-be physicists that has produced a lot of good engineers). "So then I went to aerospace because it was pretty much the coolest major there. I thought if I want to do something, it'd better be something I think I'll love doing."

Something else Jennifer has in common with many others at JPL and NASA is that as much as she loves her work, it's not the only important facet of her life. Sports continued to give her an outlet for competition and fun, and as is typical for the people who go on to impressive careers at JPL, she excelled in two disparate areas at once: in addition to earning her degree in aerospace engineering, Jennifer played volleyball for MIT, becoming team captain in her senior year and finishing with an overall record of 106 wins to 26 losses. In 1989, she was named an Academic All-America athlete and inducted into the organization's hall of fame.

But sports and scholastics still made up only part of Jennifer's life. She is a devout Christian and deeply committed to volunteer work helping others.

"After college, I wanted to do the Peace Corps for a while because I've always had an interest in that. I had spent some time in Peru for a summer when I was in college, working in a camp for homeless children. The children weren't there, but we were building the cabins—basically just hauling thatch in the mountains, pretty much physical labor. But I always enjoyed doing stuff like that and I wanted to try it again. I didn't want to look back later in life and say, you know, I wish I had done that."

So after graduating, seeking a new goal in her life, Jennifer had interviews with an eclectic assortment of potential employers: the Peace Corps, JPL, the Boeing Company, and the CIA.

"I thought I would eventually go into aerospace, but coming right out of college, I thought the Peace Corps was a great opportunity. In fact, because of what I had seen in Peru, I really felt like agriculture was something critical to teach people." As it turned out, the Peace Corps was eager to hire Jennifer because she had a technical degree and they wanted her to teach computer science. However, she recalls, "I thought that was a little less fundamental than agriculture, so I just decided not to do it."

As for the CIA, that brings on another round of laughter and a thoughtful insight into how well she knows herself. "I think I would have liked the CIA," Jennifer admits, "but I talk a lot, and I like to talk about what I do, and living a covert life for me would be something just impossible, because I'm always yapping about things I shouldn't be! So I thought it's not my personality, even though I found it kind of intriguing."

Her interview at Boeing was for a starting aerospace position, and in the end, the company did not make an offer. But JPL made up for Boeing—her interview there resulted in her receiving *two* job offers.

"I could have started in Power or in Systems, but I decided to start in Power, which is a subsystem level." But then as now, Jennifer remained committed to her volunteer work, and between graduation and starting work at JPL, she spent the summer volunteering at a Christian girls' camp in New York.

Jennifer started work at JPL in October 1990, about the same time Wes Huntress was finalizing his initial plans for his new, low-cost Discovery missions at NASA Headquarters. Though Jennifer didn't know it at the time, her future and the future of the Discovery program were already heading in the same direction.

At JPL, even the most gifted new employees have to start at the beginning—which is exactly what Jennifer did. "When I first started, I worked on all these 'concept' missions. The Lab brings new hires in and puts them on these missions that don't exist and may never exist. But they get proposed to NASA Headquarters and one out of two hundred gets funded.

"I worked on Lunar Observer—which you've never heard of because it never got built—and some other missions that never happened. But I got into the detailed technical design of power systems, which is what I wanted. I wanted the details." After a year in that position, Jennifer decided to transfer to another subsystem area. She had set herself the goal of eventually working at higher levels of the business—Systems, then Projects, then Program—but "I wanted to hop around at the detailed level for a while."

Jennifer's next subsystem assignment was in Attitude Control, which relates to the control of spacecraft orientation after launch, during cruise, and, in the case of surface landers, during entry, descent, and landing. Precise and proper positioning of a spacecraft is absolutely essential, affecting everything from the aim of its rockets and efficiency of its solar panels, to ensuring that radio antennas are pointed back at Earth and that the angle of attack of atmospheric entry is such that the spacecraft will fly safely.

As an example of the long-term planning that's required for major space missions, Jennifer's Attitude Control assignment in 1991 was for the Cassini Saturn mission, which arrived at Saturn on June 30, 2004.

For someone who thrives on new challenges and variety, the long-term aspect of the job wasn't what Jennifer had hoped for. The *Cassini* launch was still seven years off, which meant it would be that long before anything she worked on was actually used in flight. And then it would take another seven years for *Cassini* to reach Saturn.

"It just made me think—I understand this is engineering and it was a big flight project and I was this little cog in a big huge wheel—but didn't feel I could have much impact." It was then another opportunity arose that appealed to her personal and religious commitment to help others. "I took a Christmas trip to Europe with some friends to play volleyball—as part of a church mission team. We were there for a couple of weeks and as a result, I decided to take a year off and live in the Ukraine of the former Soviet Union, and be a missionary, to teach the Bible in English over there." Again, Jennifer felt strongly that if she did not take advantage of this opportunity, then in the years ahead, she might always regret having missed the chance.

She recalls that there were people at JPL who cautioned her, saying, "That's not a good career move. And I did feel a little bad, because my parents had spent their retirement sending me to a very expensive college. But I just felt that I wanted to do that for a year, so I spent a year in the Ukraine teaching in Sevastopol, which is on the Black Sea."

As for her future employment at JPL, Jennifer knew the risk she faced by taking a leave of absence. "In JPL's leave-of-absence rules, it says: 'Only if we have a job for you, do we have to give you a job when you get back.' So I was certainly not guaranteed the job I had, which I didn't really want anyway." That admission prompts more laughter, as she admits, "It was actually a good way to leave! Sorry, I'm taking a leave of absence—find someone else to do this work!"

In the end, though, Jennifer didn't have to worry about her job prospects. After a year of missionary work, she spent a few months traveling through Europe with friends, thinking about her future options. At one point in her travels, "I sent a postcard to one of the guys I know from JPL—someone who knows everything about everybody and everything—and I said I'm going to be back in the States. Any jobs available? He called me the day I got back to Ohio to see my family, and said, 'We have this job and we need to fill it in the next day. You've got to take it. It's on this Pathfinder mission.'"

## IT JUST SORT OF HAPPENED

Jennifer smiles as she remembers that fateful phone call in July of 1994. "You know, Pathfinder didn't have a 'good' reputation at that point—it never really did until it was successful. But it was kind of that reputation that drew a lot of the people to it—the idea that they've got something hard to do."

Jennifer started back to work at JPL in August and soon came to realize that Pathfinder was indeed unlike any mission the Lab had run before. Though she had been hired to be cognizant engineer for the ground software for the Attitude Information Management (AIM) subsystem, that wasn't the job she ended up doing. "The structure of Pathfinder was so open, people could basically do what they needed to do to get the job done. There were no real restrictions on what you were allowed to do. If you wanted to take on more responsibility, or if you wanted to move to another area because you said that that was an area that needed help, that was okay. It gave people flexibility.

"So for the integration and test area for AIM, we were trying to *test* the spacecraft the way we *operated* the spacecraft."

Testing, of course, is the irreducible requirement of any space mission. There is even a document called the "irreducible test list," which states if the tests it lists are not accomplished, the mission will not be given approval to launch. When hardware costing hundreds of millions of dollars is operating millions of miles from the nearest repair station, that hardware *must* be shown to perform as expected under *all* testable conditions, even if some of those conditions are never expected to occur.

This is part of the "Test Like You Fly, Fly Like You Test" strategy that is now the core principle of all spacecraft test programs.

Most projects, Jennifer explains, "test their spacecraft functionally, but not operationally." That is, as an individual component or subsystem, a camera will be functionally tested to show that it does take pictures. But in an operational test, Jennifer says, "You know the camera takes pictures, but does it take the picture at the time of day you need it to? And can you get all the images you need? And what type of resolution do you get on those images? And is the timing right for getting it down to the ground to make your science assessment and get the next set of commands back up? All those things were the types of things we were testing. So that was fun."

Though Jennifer began her work on *Pathfinder* testing by concentrating on the AIM subsystem, under the mission's flexible structure, her job expanded according to her interests and experience and the mission needs. As *Pathfinder*'s other subsystems and instruments were delivered, she took on responsibility for more test procedures. By the time the full spacecraft was coming together, she became the "lead"—that is, the person in charge—of all testing of surface operation sequences. "Essentially, I was the lucky one who got to hook everything together. Most people were more specialized, and because I was doing operations and testing and ground software, I was the one who had to make sure it all fit together and worked—at the Systems level. Because that's what needed to be done.

"And that's what brought me to be the flight director when *Pathfinder* landed because I was the one person who had sat there for what seemed like thousands of hours watching these 'stupid' sequences run. I had seen every problem that we ever had on that thing, and it was just kind of a natural place to be. That's how I got to be flight director, but it certainly wasn't the job I was hired to do. It just sort of happened."

Jennifer is quick to point out that the Pathfinder mission had several flight directors, and it was simply her "good fortune" to be the flight director on duty for landing day. Brian Muirhead, on the other hand, would like it known that Jennifer's selection as flight director was made on the basis of her knowledge and experience. She was clearly the best qualified to deal with any issues that might arise on this critical day. Also, in the weeks leading up to the landing, good fortune might not have been the first term to come to Jennifer's mind—and it all had to do with the results of all those tests.

"We were so afraid it was going to fail," Jennifer says, "because all these operational readiness tests we had—they all failed in some way or another. And some of them failed *really* bad, you know. Part of the problem with the Operational Readiness Test [called ORTs, of course] is that you have to simulate all the conditions in space in addition to testing your spacecraft, so you might have the sun in the wrong position, and that would cause part of your test to fail.

"But we did seven full-up tests with the big team there. We operate on real [Mars] time, and we essentially run through entry, descent, and landing, beginning to end. Then we get our first simulated downlink and we try to assess the health of the spacecraft and what we need to do to get our images back to Earth and to get the

## THE CROWD GOES WILD

*Public Reaction to the* Pathfinder *Landing*

Media training? What would a bunch of camera-shy engineers and scientists need media training for? Here's one good answer: to explain why, after spending hundreds of millions of taxpayer dollars, you haven't heard from your spacecraft.

One of the wonderful things about working on the Mars Pathfinder mission was that we were largely below the radar screen. JPL and NASA management knew we existed but didn't pay a lot of attention to us, until, that is, it looked as if we were actually going to make it to the launchpad. So the core team signed up for media training, except for project manager Tony Spear, who had endured it years before and insisted he didn't need a refresher.

For those of us who were new to the game, we worked in a simulated briefing room in the JPL TV studio, and we learned some of the tricks of the trade: Be honest; answer the question directly; if you don't know say so; and above all, don't speculate. We gave mock briefings with our consultants throwing tough questions at us while taping, then replaying and critiquing our performances. Surprisingly, everyone took to it pretty naturally except our software lead, Glenn Reeves. Glenn just couldn't relax enough to stop making sarcastic comments, so we took him off the list, and he thanked us.

After *Pathfinder* launched successfully, I remember arguing with the media relations people about having to allocate precious budget reserves to support media operations—providing trailers, phones, people, et cetera to work with the media on landing day. The original estimate from media relations was about $500,000, and I thought that was outrageous. That $500,000 represented four engineers who could be doing testing, doing something to contribute to mission success. (Silly me, I didn't yet understand the meaning of mission success.) But we were told we had to do it, so I negotiated them down to $300,000. Unfortunately, by the time it was all over, we spent the original $500,000 and then some—such is the price of success.

Until *Pathfinder,* my only previous experience with a media event was seeing the preparations for the *Voyager* encounters with Jupiter, Saturn, Uranus, and Neptune during the 1980s. These seemed liked big deals, lots of dignitaries, press. I found it difficult to believe that *Pathfinder* would really draw that kind of attention. But as landing day approached, media relations informed us of how many requests for accreditation they'd received, and it was in the hundreds. They said this was going to be much bigger than *Voyager,* which is why they needed more money.

Because our optimum launch opportunity had included the Fourth of July as a possible landing day, we had chosen it as a sales ploy to cement the date in the minds of decision makers. But as the day approached, we started to realize that this would be a holiday and that people might actually tune in and watch what was going on.

We had scheduled various press briefings leading up to landing day. Given all the possibilities, good and bad, we worked hard to manage expectations. Our media package even explained a few of what NASA calls "off-nominal" scenarios, in which we might not hear from the spacecraft right away, and why that didn't necessarily mean we were in trouble. The briefing I feared most was scheduled for 2:00 P.M. on landing day. This was to be the briefing to explain exactly that possible situation . . .

The morning of landing day all was quiet—as in, just before the storm. In our little mission control room without windows we really had no idea what was going on outside regarding the weather or the media. We were focused on Mars. There was one cameraman in the room with us, John Beck, who was providing a live feed to NASA TV and the media. He caught all the drama and excitement of the events leading up to entry, from how we all held our breaths during the five minutes of Entry, Descent, and Landing, to the joy and tears of receiving that signal from the surface of Mars that told us we'd made it.

In the time between landing (which was 3:00 A.M. MLT, Mars Local Time; 10:06 A.M. PDT, on Earth) and the next period of communication (after sunrise on Mars), we had time to go down to the media relations area. JPL had converted our little museum and main auditorium into a makeshift press room and news conference area. When I arrived, there was a lot of buzz about something big that was going to happen, and sure enough the once-dreaded 2:00 P.M. press conference was now scheduled to become a teleconference with me, JPL Director Dr. Ed Stone, NASA Administrator Daniel Goldin, and Vice President Al Gore. It was certainly more fun than the alternative we had trained for. We had thought much more about being prepared for an anomaly or disaster than we had for success. We just must have figured that success would take care of itself.

We had always said that we'd be considered a success if we could just get one picture from the surface of Mars on the front page of *The New York Times.* We called that our "mission success panorama" and we actually got it down and processed on the very first day. We showed it at our first major science press conference at 6:30 P.M. Pacific Time on July 4. On the stage were Daniel Goldin, NASA's Associate Administrator for Space Science, Dr. Wesley Huntress (who is the true father of Mars *Pathfinder*), Peter Smith, the camera principal investigator, Project Scientist Dr. Matt Golombek, and me. The excitement in the press room was almost as great as the excitement up in mission control after we landed. The media was living the moment with us.

On the morning after, groggy from too little sleep, I opened the front door and saw the front page of the *Los Angeles Times* with its large, full-color picture of Mars, and the headlines: "A Picture Perfect Day on Mars."

Rob Manning addresses the crowd of Mars enthusiasts at Planetfest '97, held in Pasadena, California, to celebrate the *Pathfinder* landing on the Fourth of July weekend, 1997. COURTESY THE PLANETARY SOCIETY.

I yelled, only half in jest, "We really did do it!" It wasn't a dream.

On the way in to work, I went by the local newsstand and there it was, our mission success panorama on the front page of *The New York Times*—and on every other paper I could see. Over the course of the next few days we marveled at the fact that the coverage and headlines were coming from literally all over the world. As I commented at a later press conference, when asked if we were surprised by the degree of interest, I said: "Absolutely, we had no expectation that there would be anything like this. I can only attribute it to the fact that people see this mission not as an accomplishment of JPL, or NASA, or the United States, but as a human accomplishment. And we're very proud to be a part of it."

After I got to work on the morning of July 6, I looked at my calendar and the upcoming schedule of the day's activities and suddenly realized that I had completely forgotten a commitment I'd made weeks ago to speak at

Planetfest, a celebration of space exploration that the Planetary Society was sponsoring at the Pasadena Convention Center. It was almost 12:00 P.M. and I had no material at all to present. (Engineers can't talk without PowerPoint slides.) I thought seriously about calling Lou Friedman, executive director of the Planetary Society, and telling him I couldn't make it, but I hated the idea of breaking a commitment without calling and I didn't know how to reach him. I figured that there would only be a few people there and that I could just wing it, so off I went. When I got to the convention center it seemed pretty deserted and I didn't even know which room I was supposed to be in. But Lou saw me and waved me toward a door.

The instant I stepped through that door, hundreds of people erupted into applause and cheers. As I walked to the stage, they all stood up and just kept clapping. All I could do was smile. Wow, this is how rock stars feel, I thought to myself. I apologized for not

having prepared a talk, so I offered to share some stories about how we got to where we were. They loved it, I loved it—all the stories of the challenges, crises, the great people, just came gushing out to cheers and laughs and more ovations. When I finally begged permission to leave so I could get back to Mars, I was literally mobbed trying to get out of the auditorium by adults and kids wanting to congratulate me, us. My fifteen minutes of fame had begun.

—B.M.

rover off the lander. In addition to the full-up tests, we did a lot more mini ones and we had several where we tested different contingencies."

By "contingencies," Jennifer means "disasters in the making." The Pathfinder team would try to guess what components on the spacecraft might fail, or simply suffer from a stroke of bad luck. What if a deflated airbag snagged on a lander petal and couldn't be tucked up out of the way of the rover? What if the high-gain antenna failed? Could the rover still be deployed with images transmitted only through the low-gain antenna?

"I can't remember all the contingencies," Jennifer says, recalling those hectic days. "There're so many things that can go wrong—you just can't imagine all the things that can go wrong. For some tests, we put in the one-way light-time delays, so we wouldn't get the data for ten minutes on landing day, and commands wouldn't get to the spacecraft until ten minutes after we had sent it. But there were occasions during our testing where in order to not mess up the entire test, we had to take out the light-time delays, which you can't do in the real world. I'd think, gosh, I hope that we don't have to do this when we're on Mars, because we *can't* do it."

One key contingency that the team prepared for was the nightmare situation in which "We just never heard from the spacecraft. In case that happened on *Pathfinder*, the team equipped the onboard computer with a set of programmed sequences so that if the spacecraft was operational after landing, "It would continue to try to do things, and it would continue to try to communicate with us, so if for some reason it couldn't get our commands, it could still send data back. The spacecraft might be deaf but it would continue its mission."

The week before landing, that scenario was the one Jennifer studied most. Why? "There were already a lot of media folks around," Jennifer explains, "and I knew that I would be standing there on landing day, and I was so scared that we would just not hear a thing. That the time would come and we would expect to see data and there would be no data. So I studied the procedure for everything that we would do on sol 1 [a "sol" being a Martian day] if we didn't heard from the spacecraft. Because at least I would know what to say: Okay, we didn't hear from it, but we expect the spacecraft to be operating this sequence and it will try to communicate with us again at this time and if we don't hear from it then, it will try to do this next thing."

But as it turned out, Jennifer's preparations were not required, because the *Pathfinder* landing was flawless.

When the team realized that the Entry, Descent, and Landing sequences had run perfectly, and that they had a healthy, functioning spacecraft on Mars, Jennifer recalls, "We were all kind of in a state of 'unbelief' because for four years everything we had done had been focused on this moment, and to finally get there and see it work, I think we were all astonished. We knew we had done everything we possibly could, but there are things that you can't control in the environment of Mars, there were all these things that could go wrong, and the whole world was watching."

## MOTHERS, HIDE YOUR BLENDERS

To fail on Mars is a terrible thing, because usually there are no second chances.

While a definitive explanation for the loss—more than two years after *Pathfinder*'s success—of the Mars Polar Lander in December 1999 will have to wait until the spacecraft or its wreckage is examined by future explorers on Mars, NASA's subsequent investigation revealed the likeliest scenario: the lander's descent rockets were designed to shut off when the lander's legs registered contact with the Martian surface. A problem during testing resulted in part of a test not being run in the right configuration. This resulted in engineers not seeing there was a flaw in the flight software. Engineers knew that when the legs deployed into the landing position hundreds of feet above Mars, that movement could generate the same signals that indicated touchdown. But a software design error failed to reset, before touchdown, the parameter that the computer used to trigger the engine shutdown. Thus, the review board concluded it was probable that when the Mars Polar Lander's legs moved into landing position, the onboard computer received a signal telling it the craft had landed and the rockets were shut down. The result of this? The spacecraft fell the last few hundred yards to Mars—a drop it was not designed to withstand.

When failures occur during testing, of course, engineers are able to observe the

The Entry, Descent, and Landing procedure for the ill-fated Mars Polar Lander was almost identical to that of Mars *Pathfinder*, until the stage at which the lander was cut free of its backshell and parachute. *Pathfinder* completed its journey by bouncing to a stop protected by airbags, while the Mars Polar Lander was designed to touch down gently, using rocket engines to slow its descent like the Viking landers. COURTESY NASA/JPL.

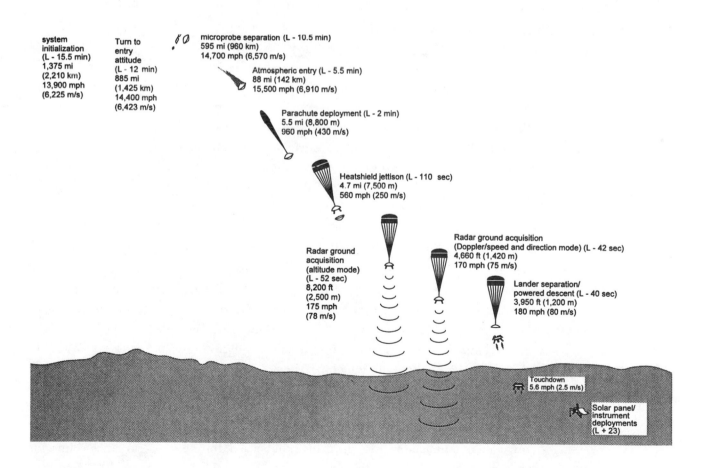

system
initialization
(L - 15.5 min)
1,375 mi
(2,210 km)
13,900 mph
(6,225 m/s)

Turn to
entry
attitude
(L - 12 min)
885 mi
(1,425 km)
14,400 mph
(6,423 m/s)

microprobe separation (L - 10.5 min)
595 mi (960 km)
14,700 mph (6,570 m/s)

Atmospheric entry (L - 5.5 min)
88 mi (142 km)
15,500 mph (6,910 m/s)

Parachute deployment (L - 2 min)
5.5 mi (8,800 m)
960 mph (430 m/s)

Heatshield jettison (L - 110 sec)
4.7 mi (7,500 m)
560 mph (250 m/s)

Radar ground acquisition
(Doppler/speed and direction mode) (L - 42 sec)
4,660 ft (1,420 m)
170 mph (75 m/s)

Radar ground
acquisition
(altitude mode)
(L - 52 sec)
8,200 ft
(2,500 m)
175 mph
(78 m/s)

Lander separation/
powered descent (L - 40 sec)
3,950 ft (1,200 m)
180 mph (80 m/s)

Touchdown
5.6 mph (2.5 m/s)

Solar panel/
instrument
deployments
(L + 23)

ovent and analyze the problem in detail to determine the cause, make corrections, and test again. And sometimes failures of the most unusual kind can lead not just to improvements, but to a totally new direction.

Which brings us to a classic story of an engineer's mind at work.

To begin, meet Tommaso Rivellini. If we compare JPL to James Bond's MI6, then Tom would be Q, the mechanically minded genius who can pack laser disintegrators into wristwatches and transform sports cars into submarines. One of his most notable designs is the complex arrangement of airbags that protected the *Pathfinder* and both MER landers—designs that include not only the size, shape, and number of layers of fabric in the bags, but the methods used to inflate them. And—more important—the interior network of cords releasing the gas from the bags when it's time to deflate, then retracting them beneath the landers so the rovers have clear access to the Martian surface.

How does anyone design something like that? Well, would you believe Tom started out with ordinary garbage bags, tape, and string? It makes perfect sense that he would use items available in almost any kitchen, because it was in his family's kitchen that his career in engineering began—as a toddler taking his mother's appliances apart.

"My mom would hide her blenders on the top shelves. Literally," Tom says. And as far as he was concerned back then, he wasn't actually taking things apart. Instead, he calls it "deconstructing before I knew how to construct. Because, usually, when I put them back together I'd have parts left over. So that was the foundation. I really had no idea what engineering was at the time."

Tom's parents were both physicians, so his first introduction to the outside mechanical world came courtesy of his cousins. "I was probably in junior high when my cousins were into formula car racing." Young Tom's reaction to aerodynamically shaped cars traveling at high speed? "Wow, these are cool!"

"In retrospect," Tom says, "it seems that I was quite naïve not knowing what engineering was until the ninth grade, or something like that. I mean I knew what science was. I was definitely into technology. I loved model airplanes, cars, and boats. I was building all those kind of things."

But formula racing continued to be a major interest, and by the time he was in tenth grade, Tom was no longer just playing with his formula racing car models, he became involved in a hands-on way—by building his own version of a wind tunnel. "I would see the guys in white coats with those smoke generators studying the way the smoke would stream over the cars' aerodynamic shapes. So I took an old cardboard box, put little windows in it and took my little Formula One model cars and stuck them in there with a fan. For smoke, I bought an incense stick and basically got totally dizzy because I didn't open the window. And sat there for hours watching the smoke go all over the cars."

Fortunately, he survived his dizziness, but he was hooked. At the time, he was attending high school in Gouverneur, New York, a small agricultural and mining town about a half hour from the Canadian border. Tom told his guidance counselor that he wanted to be an aerodynamics engineer, and the counselor deftly steered him into aerospace engineering. "I was really enthusiastic, so I started looking for colleges."

Because he wanted to stay close to home, Tom attended Syracuse University to study aerospace engineering. But his focus was on aircraft and cars—the idea of spacecraft and space exploration were still in the future, though not out of the question.

"I was really into airplanes, but I don't recall saying, 'Oh, I definitely want to design airplanes or cars.' But my curriculum was pretty much focused on aeronautics. The first couple of years of aircraft engineering are very basic, and it's in the last two years where you start to specialize. Even then, the specialties are typically very general." But not always. Tom's personal trajectory took a sharp turn toward Mars in his senior year, 1989, when one of the elective classes he took was Spacecraft Dynamics. Once again, the favorite word of most JPL engineers and scientists comes into play: "I thought, 'This is pretty cool.'"

Tom recalls that most of his instructors were oriented toward straight academics and research. However, "There were a few exceptions where there were some teachers who gave us a strong dose of reality." He remembers that design classes were a highlight, particularly because of the real-world experience they offered. "In some class they would say, 'This is how you design a spring, and here're the mathematical equations that describe two springs interacting together.' And a spring could be a rocket, an airplane, anything else that behaves in a certain way. So it was very analytical, yet practical. Most students really enjoyed the design classes because you really got your hands dirty and it was fun."

Tom enjoyed the hands-on "fun" so much, he had no doubt about going on to graduate school. "So I had to decide what area I really wanted to go into. I got really interested in spacecraft, and in understanding more about how spacecraft worked. I wasn't sure what specific area I was interested in, just that that's the general area I wanted to be in. So I looked around—and there are even fewer schools offering orbital mechanics and spacecraft design than aeronautical engineering. The University of Texas at Austin had the most general program. It had a heavy dose of theory, but it was also the only one that offered a spacecraft mission design, along with spacecraft design. So that was very appealing to me because it had both the theoretical and the practical."

As it turned out, the practical aspect of spacecraft design was where Tom's passion was taking him. "It was a totally new field for me because I went from aerodynamics and fluids and very tangible things like that, to orbital dynamics, which is much more abstract and mathematically intensive. Fortunately, I chose the school I did because it gave me exposure to both the practical requirements of mission design, spacecraft design, and subsystem design, as well as mechanics and theoretical dynamics. And I very quickly realized that I wasn't too excited with the highly theoretical, abstract aspects, though it was good that I was learning them. But I was really excited by the 'down-to-earth' elements of spacecraft design."

Then as now, NASA was the primary customer when it came to spacecraft design, and Tom quickly made contact. "Working on my thesis, I went to the Johnson

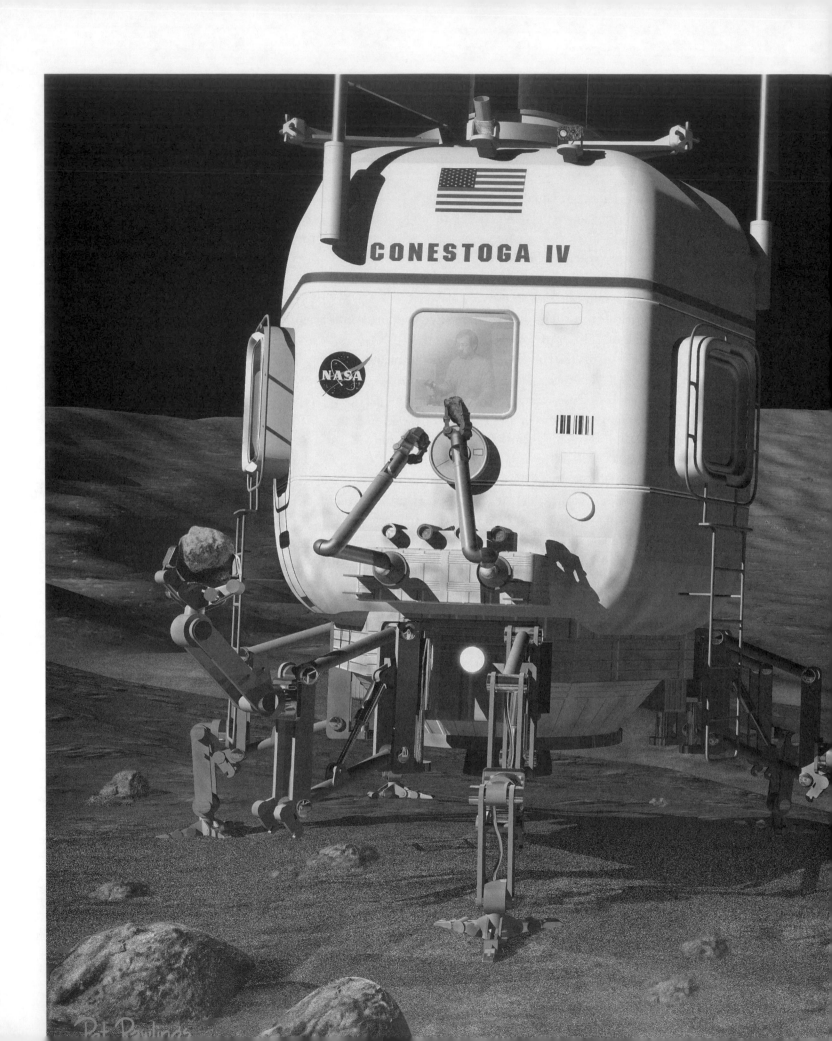

NASA is always planning for the future. This illustration shows a lunar base concept from 2003. COURTESY NASA/JSC.

Space Center and met a lot of people there." Tom's thesis was a study of lunar-based transportation, involving a hypothetical lunar base. "It was very much a vehicle for me to learn whatever I wanted to learn. But just talking to the people at Johnson Space Center, seeing the kind of things that people were actually working on, was really exciting."

Finally, it was time for Tom to have his own run-in with random chance. Or, as he puts it, "This is when, I guess, destiny reared its head. When I was about to finish grad school and start looking for a job, I wrote out my first résumé. One format for a résumé is to write down the position you're seeking and describe what you're looking for. I wrote down, 'advanced spacecraft design.' That sounds really neat, right? But after a couple of weeks, I thought, This sounds pretty hokey. I'd better get rid of that and put something much more conservative."

In one sense, choosing a conservative route was probably a good idea. At the time Tom was graduating, he remembers that "It was a bad time in the aerospace industry, and I spent a long time looking." In addition, Tom was still more comfortable closer to home. "My family was on the East Coast. I was from the East Coast. I wanted to stay on the East Coast. I said to myself, I've been looking for jobs everywhere but California, because I don't want to be in California. I actually said that."

However, though Tom came close on a few East Coast jobs, all of them fell through. "Then," he recalls, "my advisor finally said, 'I know somebody who works at JPL. You might want to give that a shot.' So I started calling around, first one person, and the next, and the next. And finally I got a callback from one of the administrative assistants and he says, 'We received your résumé and we've got a couple of supervisors who are looking for people. One of them is Don Bickler.' I say, 'What does he do?' The assistant said, 'Well, he works on Mars Rovers and spacecraft designs, and structures.'"

Tom's interest level rose immediately, and he tried to remain calm as he replied, " 'Really, what's the name of his group?' To which the assistant answered, 'The Advanced Spacecraft Development Group.'"

Bingo.

Back in 1984, that same group had been called Mechanical Support. But when Brian Muirhead had joined it as group supervisor, one of his first acts had been to change the name. And it was that name change that matched so closely Tom Rivellini's career goal—one that he had felt compelled to edit on his résumé because he thought it was "pretty hokey."

Tom's answer to this news was an enthusiastic, "Okay! I think I would definitely be interested!" A few days later, Tom received a phone call from Don Bickler, which led to an interview at JPL. And even though the Lab was the full width of the country away, "The stuff that they were doing here, it was exactly what I wanted to do," Tom says. "So they made me an offer and I said, 'I'm coming!'"

The year was 1991. MESUR Pathfinder was moving forward but not yet approved as a flight project, and the next Mars mission—the Mars Observer—was still about a year from launching. So, with no Mars flight projects really active, Tom went to work on the rover technology demonstrator dubbed Rocky IV.

One of the first rovers Tom Rivellini worked on—*Rocky IV.* COURTESY NASA.

Working with Randy Lindemann, Tom was able to design a piece of hardware for the rover, "So that was real exciting for me." His next foray into flight hardware was to design some brackets to hold an antenna in place on a defense satellite. A minor assignment, to be sure. "But it was going to go up in space! And it was really cool!"

After that, Tom's workload exploded and he became involved in multiple studies for different advanced projects—so many that he can't remember them all today. However, one study that came to his attention was Tony Spear's MESUR Pathfinder. Tom immediately went to his supervisor, Don, and told him he wanted to be assigned to Pathfinder, and Don made that possible. "So I started working on that and unlike the other studies that ramp up, you give them a design, and then they say

## "THE BEST-LAID PLANS . . ."

*Milestones of the Recommended U.S. Space-flight Program*

Because of the challenges, cost, and complexity of space missions, NASA must constantly look ahead, its scientists and engineers doing their best to set future goals based on the knowledge of the day. For example, the suite of scientific instruments carried by the Mars Exploration Rovers of 2004 grew out of initial studies and development programs that began in 1984.

Back in 1958, the year the National Advisory Committee for Aeronautics (NACA) was absorbed into the newly formed National Aeronautics and Space Administration (NASA), that drive to look to the future was already in place.

This is how the future looked in July of that year. We've added a column to the right to show how that future unfolded.

From NACA, Special Committee on Space Technology, Working Group on Vehicular Program, "A National Integrated Missile and Space Vehicle Development Program," July 18, 1958.

| ITEM | DATE | PLANNED EVENT | SCORECARD |
|---|---|---|---|
| 1 | January 1958 | First 20-pound [Earth-orbiting] satellite | **October 4, 1957**<br>The 184-pound Russian satellite *Sputnik* becomes the first to orbit the Earth.<br>**January 31, 1958**<br>The 31-pound *Explorer 1* becomes the first U.S. satellite to achieve orbit. |
| 2 | August 1958 | First 30-pound lunar probe | **March 3, 1959**<br>NASA's 13.5-pound *Pioneer 4* passes within 37,500 miles of the Moon as planned. Two months earlier, when it failed to impact the lunar surface, the Russian *Luna 1* passed by the Moon. (In other words, it missed!) |
| 3 | November 1958 | First recoverable 300-pound satellite | **August 11, 1960**<br>A 1,873-pound prototype of a U.S. top-secret Corona spy satellite, given the public cover name *Discoverer 13,* becomes the first object to be recovered from orbit. |
| 4 | June 1959 | First powered flight with X-15 [experimental rocket-powered aircraft] | **September 17, 1959**<br>They were close on this one. On June 8, 1959, the X-15 had its first unpowered test flight. Its first powered flight took place about three months later on September 17. |
| 5 | November 1959 | First 400-pound lunar probe | **September 13, 1959**<br>The 853-pound Russian spacecraft *Luna 3* becomes the first probe to hit the Moon. |
| 6 | December 1959 | First 100-pound lunar soft landing | **January 31, 1966**<br>The 3,482-pound Russian probe *Luna 6* achieves the first soft landing on the Moon (using a simple airbag system to cushion the shock of impact).<br>**June 2, 1966**<br>NASA's 644-pound *Surveyor 1* makes the first successful U.S. landing. |
| 7 | January 1960 | First 300-pound lunar satellite | **April 3, 1966**<br>The 3,520-pound Russian probe *Luna 10* became the first spacecraft to achieve lunar orbit. |
| 8 | July 1960 | First wingless manned orbital return flight | **April 12, 1961**<br>Russian cosmonaut Yuri Gagarin becomes the first human to orbit the Earth.<br>**February 20, 1962**<br>Astronaut John Glenn becomes the first American to orbit the Earth. |

| ITEM | DATE | PLANNED EVENT | SCORECARD |
|------|------|---------------|-----------|
| 9 | April 1961 | First 2,500-pound planetary or solar probe | **December 14, 1962**<br>NASA's 447-pound *Mariner 2* probe flies past Venus and returns scientific information.<br>(The Russian probe *Venera 1* had flown past Venus on May 19, 1961, but since its radio had failed shortly after launch, no information was returned and the mission was not considered a success.) |
| 10 | August 1962 | First winged orbital return flight | **April 12, 1981**<br>First flight of the Space Shuttle *Columbia*. |
| 11 | November 1962 | Four-man experimental space station | **June 7, 1971**<br>The Russian station *Salyut 1* begins operations with a two-person crew.<br>**May 25, 1973**<br>The U.S. station *Skylab* begins operations with a three-man crew.<br>**January 12, 1978**<br>The Russian station *Salyut 6* becomes the first station to have a crew of four. |
| 12 | February 1963 | First 3,500-pound unmanned lunar circumnavigation and return | **September 15, 1968**<br>Russian *Zond 5* circumnavigates the Moon and is recovered from the Indian Ocean six days later. |
| 13 | July 1964 | First 3,500-pound manned lunar circumnavigation and return | **December 21, 1968**<br>The three-man crew of NASA's *Apollo 8* mission become the first to orbit the Moon and return. |
| 14 | September 1964 | Establishment of a 20-man space station | **2010**<br>When completed, the International Space Station is designed to accommodate a full-time crew of six. Continuous occupation of the ISS began on November 2, 2000. |
| 15 | July 1965 | Final assembly of first 1,000-ton lunar landing vehicle (emergency manned lunar landing capability) | The possibility of having a backup lunar-landing vehicle for a Moon mission was not pursued past 1966. |
| 16 | August 1966 | Final assembly of a second 1,000-ton landing vehicle and first expedition to Moon | **July 20, 1969**<br>NASA's *Apollo 11* astronauts Neil Armstrong and Buzz Aldrin become the first humans to land on the Moon. Their lunar module, fully fueled, had a mass of 16.7 tons. |
| 17 | January 1967 | First 5,000-pound Martian probe | **July 20, 1976**<br>The first of NASA's two *Viking* landers successfully touches down on Mars. Each lander arrived in orbit of Mars docked to an orbiter. The total combined launch weight of each lander/orbiter spacecraft was 6,590 pounds. |
| 18 | September 1967 | Completion of 50-man, 500-ton permanent space station | **2010**<br>When completed, the International Space Station will have a mass of about 500 tons, but will support a crew of only six. |
| 19 | 1972 | Large scientific Moon expedition | **2014**<br>Under NASA's 2004 Vision for Space Exploration, NASA will resume lunar expeditions around the year 2014. |
| 20 | 1973/1974 | Establishment of permanent Moon base | **2020**<br>Under NASA's 2004 Vision for Space Exploration, a permanent lunar research station is to be established around the year 2020. |
| 21 | 1977 | First manned expedition to a planet | **2020 and beyond**<br>When this chart was drawn up in 1958, scientists thought Venus was a reasonable destination for a human expedition. However, for now, with foreseeable technology, Mars is the only other planet in our solar system on which humans could land and survive in spacesuits. |
| 22 | 1980 | Second manned expedition to a planet | **2020 and beyond** |

thank you and go away, this one started picking up steam. That's the one that turned into Mars Pathfinder. That's how I got started."

Tom's first assignment on *Pathfinder* was to come up with a "simple" way to land that didn't require powerful rockets to decelerate the lander, as was done on *Viking*. With one coworker, Mike O'Neal, and two senior engineers, Bob Bamford and Jim Hendrickson (who had technically retired from the Lab but couldn't resist the challenge), Tom recalls, "We brainstormed dozens and dozens of ideas and, finally, came up with using airbags to protect a tetrahedron-shape lander on the inside. Tom smiles as he remembers *Pathfinder*'s early days. "You know, we thought we were breaking new ground, totally inventing new stuff. And then, sure enough, somebody throws on my desk a paper from 1967—the year I was born—and it turns out this guy down the hall, Bill Layman, worked on an airbag project for landing things on other planets. It never got very far back then, but the concepts that they put forward, well, ultimately, we took what he had done back then and what we were coming up with now, put it together and that's how we came up with the Mars Pathfinder landing system."

The system ultimately worked as designed. And because it's now qualified as tried-and-trusted "flown" hardware, when the Mars Exploration Rover mission was proposed, the initial idea was to basically rebuild two *Pathfinder* landers to carry two new rovers to Mars, taking advantage of the optimum launch opportunity provided by the close approach of Mars and Earth in 2003.

Time was of the essence. Originally, NASA had planned that the ill-fated Mars Polar Lander (MPL), which was to have landed on Mars in December 1999, would be followed by a similar lander—the 2001 Mars *Surveyor*. *Surveyor* would be based on the MPL design—a three-legged platform that would land by parachute and rockets. Like the MPL, the *Surveyor* could not move once it landed. However, unlike MPL, *Surveyor* would carry a microrover—*Sojourner*'s refurbished "flight spare," christened *Marie Curie*. The lander's robotic arm would act as a crane to lower the rover to the surface, in an operation that Jennifer Harris Trosper likened to the arcade game in which players try to grab stuffed toys with mechanical pincers.

However, because the loss of the Mars Polar Lander was considered to have been caused by a flaw in the landing system, NASA canceled the launch of the ready-to-go *Surveyor* lander. (Because the flaw in the landing system is understood and has been corrected, the *Surveyor* lander—now upgraded—will at last fly as the appropriately named Mars *Phoenix* Lander, one of NASA's new Mars Exploration Program Scout missions. The *Phoenix* will land in the north polar region of Mars in April 2008. It will not carry a rover, but its robotic arm will be able to dig into the Martian ground up to three feet deep to look for, among other things, ice containing complex organic molecules that, if found, could indicate the presence of biological processes.)

The twin Mars Exploration Rover (MER) mission quickly took on a new life and new design of its own. As we'll find out in the chapters ahead, the idea to "simply" rebuild and refly two new *Pathfinders* with more capable rovers quickly be-

The Mars *Surveyor* intended for launch in 2001 was patterned after the Mars Polar Lander. However, the *Surveyor* would carry the flight spare of *Pathfinder*'s rover, which it would lower to the Martian surface with its robotic arm. The mission was canceled early in 2000, after the December 1999 loss of the Mars Polar Lander, but most of its hardware (minus the rover) will fly again in 2007 as the aptly named Mars Phoenix Lander. COURTESY NASA/JPL.

came impossible, resulting in a design that looks and flies like *Pathfinder*, followed the same Entry, Descent, and Landing (EDL) profile as *Pathfinder*, but turned out to be different in almost every detail.

Naturally, Tom Rivellini could not resist the challenge to revisit the *Pathfinder* technology for EDL, which brings us to that illuminating story that takes us into the world of engineering, to show how when things go wrong, the innovative minds of engineers can create, recognize, and grasp new opportunities.

## THE GREAT DESCENT-RATE LIMITER DISASTER (ALMOST)

Tom Rivellini describes the Descent Rate Limiter (DRL) as just "the little mechanism that lowers the lander down from the backshell, while we're coming through the atmosphere."

As described back in chapter one, one of the stages of Entry, Descent, and Landing, is to get the lander away from the rockets that will fire in the final few seconds before landing. The one thing the airbags aren't designed to deal with is being blasted by rocket flame!

So, hidden in one of the lander's petals is essentially a winch that lowers the lander from the backshell to a safe distance before the airbags inflate—about seventy feet. Since simply dropping the lander to the end of a seventy-foot tape would subject it to a sudden shock that would put enormous stress on the lander and the backshell connection points, the winch is designed to play out the tape at a slow, smooth, uniform rate. Thus the name Descent Rate Limiter, because it limits the rate of descent. The heritage of this critical device is very humble: It's used by airplane pilots for emergency exit from a cockpit window in the event there's no other way to leave the aircraft.

Tom was asked to become involved in the DRL because something had gone very wrong with this "simple" device. Tom recalls that the device was a little bit behind schedule, and some important tests hadn't yet taken place. His task was to step in to make sure the tests were carried out and that nothing had been overlooked.

To Tom, it was a simple assignment. "Basically," he explains, "it's a glorified tape measure. Think of it that way. It's got a big long spool of metallic tape, which is just like a tape measure, and on the side of it, there's a little braking element that, as you pull the stuff out, engages the brake, and it prevents you from pulling it out too fast."

To be more technical, it's what engineers call a "centrifugal brake." The faster the tape tries to unwind, the harder the brake applies, so with those two actions in balance, the tape will unwind with a constant speed.

"The problem," Tom says, "was that in order for this tape to get out of the lan-

After *Spirit*'s successful landing, the exact information obtained about its speed, orientation, and movements was applied to the existing detailed computer animation of the Entry, Descent, and Landing process. The result was a new computer animation, in which every movement was a precise reproduction of what the spacecraft had actually done. In this frame, the Transverse Impulse Rocket System (TIRS) is firing laterally while the main retro rockets fired down to slow the lander's descent. The procedure worked so well that when the lander was twenty-three feet above the surface of Mars, it was virtually motionless. At that point, the lander was cut free from its backshell and parachute and fell the final twenty-seven feet, eleven inches, to Mars. COURTESY NASA/JPL.

der, it had to go through some bends and turns, and that would cause the tape to deform. The deformation would stretch the material to a point where it wouldn't go back to its original shape [called 'yielding' by engineers], and, metallurgically, that actually makes the tape stronger, because you're toughening the material [which engineers call 'cold working'].

The designers of the Descent Rate Limiter (DRL) counted on that strengthening deformation to give the tape "margin." *Margin* is a favorite word among engineers. It refers to what most people would call a safety margin—that is, extra strength or capability that gives a device or a system some room to operate beyond what it's designed to do. A large safety margin is one of the things that allows airplanes to continue to operate well after their original design life has passed.

With Tom involved in the DRL, the first tests soon took place at China Lake, using metal tape from what was called the "flight batch"—that is, a big spool from which the tape actually to be used on Mars would be cut—provided it passed the tests.

But, Tom recalls, "What happened was, when we started testing under somewhat higher loads than had ever been tested before, the tape was coming out nice and straight, but then it would suddenly break. We couldn't figure out what the heck was going on. We even checked the material order and it was the same material that had been ordered and tested before. So we were scratching our heads, until finally I said, 'Let's get the stuff looked at under a microscope and do some metallurgy on it.'"

Tom smiles as he remembers that was the decision that cleared up the mystery. The DRL team had ordered what they wanted and had received the certification paperwork from the vendor, but had never thought they needed to double-check something as basic as the metal properties.

Sure enough, metallurgical testing showed "it was just the wrong material. I mean, chemically, it was the right metal, but it was the wrong temper. [*Temper* is a metallurgical term referring to degree of hardness.] That meant the tape was too brittle to begin with. It wasn't behaving right. In fact, it was going beyond its capability and failing, because it couldn't stretch that extra little bit. It took a long time to figure that out."

As often happens in complex, interdependent systems, resolution of one problem led to another. Though the tape was coming out of the DRL at a constant speed, the interior spool unrolled faster and faster, meaning the braking force exerted on it continually increased. "So the brake pads are doing more work at the end than they are at the beginning."

Tom provides an analogy. "It's like, if you're driving down a hill, you don't want to wait until you're at the end of the hill to slam on your brakes, because you'll destroy your brakes. You want to brake evenly all the way down. Same thing with locomotives. If those guys are going down an incline, there's a certain point if they don't start braking, they're never going to stop. They've got to brake all the way down."

With the defect in the hardness of the metal tape resolved, the brake elements in the DRL were extremely close to their maximum capability, even with margin.

"In fact," Tom recalls, "there were three brake elements, and in the tests that we were doing, there were supposed to be two brake elements engaged. But in one of the tests, one of them accidentally became disengaged, so all of the braking was happening on *one* of the brake elements."

Tom shakes his head as he relates what happened next in this fateful test. "We had a camera running on it during the test, and we'd never seen this before—you're *never* supposed to see this kind of thing. In fact, we didn't think there was any physical way for this to happen, but the brake literally *exploded.*" Tom starts to laugh as he remembers the amazement he and the testing team felt. "There were *flames* shooting out of this thing, and there was nothing combustible in there! We said, 'Whoa, we're really close to the hairy edge with this thing.'" In fact, they were beyond it.

The laughter comes only in retrospect, though, because at the time the situation was grim. "I was driving home that day thinking to myself: These brake elements are up against the limits. The steel tape is up against the limit. It's going to take us six to eight weeks to get in the new steel ribbons for this next test to happen. But we were already late for delivery to flight integration, so there was no way we could afford to just sit around and wait for the material to show up, especially if we find out that even the right material isn't going to work. We've got to think of something different. We've got to think of some way to be able to use Kevlar rope—or some other kind of really high-strength rope—and not worrying about it bending and wrapping around and doing this kind of stuff." Kevlar, of course, is the exceptionally strong fiber made by DuPont that's used, among other things, for bullet-resistant vests. Tom was familiar with the material because the cords within the *Pathfinder* airbag system were made of Kevlar.

And then inspiration struck.

"It came to me that if you take a big spool of rope and you wind it up, and you start pulling on it with hundreds and hundreds—close to a thousand—pounds of force, it'll fail because that top cord will slide into and bury itself underneath those lower layers on the spool. It just gets tangled up and doesn't work. So, I'm thinking to myself: How the hell do I get the spool stored under *low* tension and somehow have it engage those brakes that we still have? Because I didn't want to redesign the whole thing. I wanted to still engage those brakes. I'm racking my brain and then it hits me: *sailboats*!"

From a craft that sails the sea to a craft that sails through space, Tom Rivellini had found an important point of technological convergence. "Sailboats have these things called a capstan winch. It's literally this hourglass-shaped spool that sits up on the top deck. You wrap the rope [the sheet] tied to the sails around it a couple of times, pull it tight, turn it on, and it pulls in the sail at a constant rate."

On his drive home from the disastrous test, Tom called all his friends who were sailors, asking them how capstans and other sailing winches work. Then, for the next two days, he did what millions of others of people do when seeking information—he surfed the Internet, searching for people who make custom winches.

"Finally, I called some guy completely out of the blue and said, 'I'm trying to do this thing for a spacecraft and you guys do winches. How does a capstan winch work?'" More laughter as he reflects on the seeming absurdity of that call. But the fellow on the other end took Tom's request in stride.

"He was really nice, and he told me, 'The problem you're encountering is called the overriding loop problem. We've encountered that here, and the way you deal with it is you have to have a secondary idler, a secondary roller.'"

The secondary roller made perfect sense to Tom. "The next day, literally, I drove back out to China Lake and said, 'Guys, we've got to do this. I've got to have a secondary spool that I can control the tension on.' So I asked, 'Have you got any ideas on how we might do the storage spool and keep tension on it?'"

The answer from his team was as unexpected as it was simple. "Why don't you go to the bike store and get a disk brake for a mountain bike? We can have our technicians weld it up, and then we can use that to store your low-tension spool and run it through our existing brake devices."

Two days later, Tom was at a bike shop, bought the parts, and had them welded together for a new test—all this for a spacecraft that was to land on Mars.

Unfortunately, the new test was a failure, too.

The thrill of inspiration was now eclipsed by the rush of panic.

"It was like, 'Oh boy, *now* what do we do?' So that's when I drove home again, got on the Internet, found some random guy and some random winch-manufacturing company. Again, he was a really nice guy, and he explained to me, 'This is what's going on, and you can't do it with one shaft. You had to do it with two shafts.'" In hindsight, after examining the reasons for the failed tests, the solution was obvious. So obvious, in fact, that Tom says, "As I'm talking to this guy on the phone in one corner of my office, and he's explaining to me that I need two shafts, my supervisor, Don Bickler, is in the other corner, drawing two shafts and saying, 'This is what you need to do.'"

At first, Tom recalls, he didn't want to go to two shafts because it seemed complicated. Complexity can mean greater risk of failure and engineers abhor failure as much as nature abhors a vacuum. But time was running out. "So we went back to China Lake with our setup modified with the second shaft and ran it. And it worked. It worked really really well."

But that wasn't the end of it.

The makeshift device they had tested was too big and bulky to fit on an MER lander. "So then we spent the next four to six weeks in a continuous, iterative cycle of: you build something, test it, figure out what worked and what didn't, and then start over."

What sort of challenge were they facing? Well, they had a working device the size of a microwave oven, and all they had to do was shrink it to the size of a casserole dish while leaving it with the same capabilities.

Also, when it came to a final flight hardware version of the Descent Rate Limiter, mountain-bike parts weren't going to cut it. Once the design was finalized, custom parts capable of surviving space were an absolute requirement. That raised another round of challenges.

## NATIONAL RADIO SILENCE DAY

*Listening for Mars in 1924*

In August 1924, Mars made its closest approach to Earth in the twentieth century. This time, not only were astronomers looking at Mars with every telescope at their disposal, a new breed of scientist was *listening* to Mars with the technological marvel of the age—radio.

At the time, the majority consensus of the scientific community was that though Mars probably was home to some form of vegetation or other simple life-forms, it likely did not harbor a civilization of intelligent beings. However, even the slim possibility that it *might* have Martians was enough for the world to cooperate in a fascinating, if little known, experiment.

The experiment was organized by Dr. David Deck Todd, professor emeritus of astronomy at Amherst College. With the cooperation of the U.S. Army and U.S. Navy, the Department of Commerce, radio giant RCA, and the embassies of Italy, Cuba, and Argentina, Todd was able to have almost every major radio transmitter around the world—including public radio stations—stop broadcasting for five minutes at the top of every hour during Mars's closest approach. The idea was to eliminate as much radio interference as possible so that any signal sent by the Martians would be more easily detected. To that end, under official orders from the chief of Naval Operations and the War Department, all U.S. Navy and Army radio stations around the world were instructed to "listen-in" for possible signals from Mars, on the night of August 21–22, from midnight to 8:00 A.M. local time.

In reporting the story of Dr. Todd's request, *The New York Times* noted, "Although officials were strongly skeptical as to success, they seemed to take the attitude that there could be

"Usually," Tom explains, "when you want anything custom, no matter how simple a mechanism it is, it takes six months. You want a custom ball bearing? That's a year. You want a custom motor? That's eighteen months, minimum. You want a custom whatever? Months to years, right?"

So he made his first call to a possible supplier, took a deep breath, and explained what he wanted: a reasonably intricate device in an extremely small package. Tom asked, "Can you guys make a custom one for me? And the guy says, 'Yeah, sure, we can do it.' And I'm thinking, All right, it's going to cost me a million dollars and six months." But Tom had to ask, so he did: "How long do you think it'll take? And how much do you think it would cost?"

The prompt answer was, "I could get you something in three or four weeks, for maybe twenty thousand dollars."

Tom nearly explodes with laughter as he shouts his reply: "SOLD!"

One more technical problem remained. The Descent Rate Limiter was a friction-dependent device. The friction between the cord and that shaft was extremely important because it determined how much "back tension" was required. "If you needed a lot of back tension," Tom explains, "because you have a slippery shaft, then you're going to wind up with this big, huge tensioner [like the device that keeps car seatbelts snug while still permitting slow movements], and all of a sudden, you've got the whole high-tension problem all over again."

Since Tom wanted to use the lowest-tension spool he could make, he needed the highest friction possible between the cord and the shaft. "So I went to the machine shop and I ask, 'Can we knurl this?'" Knurling is a fine crosshatch cutting that gives roughness to surfaces like the handles of metal tools, providing extra friction and thus a better grip.

But the machine-shop technicians didn't think knurling was the best solution. They suggested Tom have the shaft treated with a new kind of plasma coating that provided even more friction. Tom had never heard of the process, but the technicians showed him a sample, and the next day Tom was at a plasma-coating supplier with his DRL parts, asking if he could have them back by the end of the day—and he could.

Finally, all the flight-ready parts were in hand for assembly and final testing. The DRL had been completely redesigned yet still fit perfectly within the space allowed for the original version, even down to the bolt holes that were used to attach it to the lander.

And this is where Tom says, "I just have to brag. I'm sorry, this is too much fun."

With the new design, "We found that by going through a separate spool onto a constant-diameter shaft, all of a sudden the descent rate was constant. It wasn't good, it wasn't bad, it was just different." In other words, the brakes were no longer operating by increasing the force they exerted over time—they maintained an even pressure instead.

"But what we also found was that, by virtue of braking at a constant rate, we were being *much* more gentle on the brake pads. So we said, 'Hey, why don't we up the ante here?' Instead of doing the drop tests at the required load, which is seven hundred and fifty pounds, I said, 'I want to go to eight hundred and fifty pounds, and we can probably do it.'"

In tests, it's not unusual to push the limits of the item being tested, if only to determine what its margin is. When possible, some tests are conducted "to destruction," in order to find out exactly what it takes to get something to fail.

"So," Tom says, the grin never leaving his face, "we did eight hundred and fifty pounds. No problem. So we said, 'Why don't we do nine hundred pounds?' So we did nine hundred pounds.

"Finally, in preparing for the flight qualification test, we had a whole bunch of the units that we tested. So we said, 'Well, why don't we go to a thousand pounds, just for the hell of it?' So we did a thousand pounds. And, you know, we're still very conservative around JPL, and some engineers said it's great that you did it once, but it'd really be great if you could do it *twice* at a thousand pounds—on the same system *without* any refurbishment."

Tom had no trouble with the request. The new DRL was already exceeding its design requirements. "What have we got to lose, right?"

(Just as a note to any aerospace engineers who might be cringing at this point. No one at JPL would be so cavalier about actual flight qualification, when the hardware design that will fly on the mission is given its final round of verification tests. Tom was working with units called engineering models whose purpose is to validate a design before committing to the final design or, in this case, the final tests.)

Tom decided to test two of the DRLs at a thousand pounds without any refurbishment. "The second time we did the test, I noticed that the thing [lander] came down a little bit fast. And I thought, Hmmm, that's odd. But it was within the time limits to come down. It all worked really well.

"When everybody left, I thought, Now I'll figure out what's going on here. I took it apart: I had installed one of the brake elements *backwards*!"

Tom can't contain his laughter. Not only had his new design exceeded specifications, it had done so with only *half* of its braking capability. Under the same conditions, the old design had exploded.


## THE MORAL

The only way to avoid risk in space exploration is to stop exploring space. Fortunately, there's another strategy that can be just as effective—*managing* risk.

As test lead for *Pathfinder*, Jennifer Harris Trosper watched her spacecraft fail tests time after time. But because those failures were in the test bed, where they could be analyzed and corrective measures could be devised, the actual spacecraft that went to Mars completed its mission with honors.

Tom Rivellini saw a key MER lander component fail spectacularly, realized that the design was at its breaking point without actually knowing why. And with no time to correct the design, he developed a completely new design that was inherently better and much more robust.

no objection to giving communication with Mars a fair trial under the best possible circumstances."

Interestingly, that skepticism wasn't necessarily based on the belief that we wouldn't hear from Martians because they didn't exist. Radio was still in its infancy, and the *Times* also reported that the chief of the Radio Laboratory of the Bureau of Standards stated that no radio signals from Mars would reach the Earth because those signals could not penetrate the "heavy atmospheric shield surrounding the Earth."

In other words, it wasn't the idea of Martians that was impossible, it was the idea that radio signals could actually travel between Mars and Earth that was in question.

Disappointingly, despite many claims to the contrary from around the world, none of the mysterious dots and dashes heard during that opposition turned out to have come from Mars.

However, as an example of how some things never change, a clever device built by Francis Jenkins and operated by the U.S. Naval Observatory, turned those mysterious radio signals into flashes of light that were then recorded on a long roll of photographic paper. When the roll was developed—all thirty feet of it—the signals appeared as what the *Times* called "a fairly regular arrangement of dots and dashes along one side."

However, the story goes on to say that along the *other* side of the six-inch-wide roll, "at almost evenly spaced intervals are curiously jumbled groups each taking the form of a crudely drawn face."

A face on Mars? In the words of inventor Jenkins, "It's a freak we can't explain."

Just like Percival Lowell in 1908, and those convinced there's another face up there today, people looked to Mars in 1924, and once again saw exactly what they hoped to see.

Sometimes, unfortunately, not all the failures occur on the ground. The loss of the Mars Polar Lander and the Mars Climate Orbiter were painful reminders of the unforgiving nature of space exploration.

But the engineering expertise developed by the teams who designed and built those spacecraft continues to pay dividends to NASA and to the aerospace community, especially since NASA studied those failures and took measures to ensure that similar failures will not happen again. One of the most significant results of the Mars '98 failures was for JPL to capture its best practices in a set of "Design Principles." The document outlining these principles, authored primarily by Matt Landeno at the direction of Tom Gavin, Associate Director for Flight Projects and Mission Success, contains guidelines for every aspect of deep-space spacecraft design, from electronics and propulsion subsystems to mass and power margins. Every project must now study this document and apply each principle or provide good engineering justification for not doing so. A tailored set of the principles is now the foundation for the design of each JPL spacecraft and instrument.

Accepting a manageable level of risk is very different from taking a pure, just-roll-the-dice gamble, but as the stories of Jennifer and Tom have shown us, *anticipating* problems is essential to the successful exploration of space. Understanding, managing, and ultimately *accepting* a reasonable level of risk makes space exploration possible and affordable. And honestly dealing with the causes and consequences of failures when they inevitably occur in this inherently risky endeavor makes the next round of exploration all the more likely to succeed.

Going back to Tom Rivellini's discovery that his DRL had performed spectacularly even with only one of two brakes, in addition to having a good laugh, Tom also brought considerable peace of mind to the MER team. "Personally, that gave me the confidence that this thing was bulletproof. It was just rock solid."

He was right.

When things go wrong, good engineers get going and get it right.

# CHAPTER FIVE

# "A Lot to Worry About"

## Pathfinder Tales

*Two British engineers have developed a plan called "The Martian Probe." They propose a three-step rocket—one that would drop two of its sections, one at a time, as their fuel load was exhausted along the way ... The Probe would scan the surface of Mars in strips about 200 miles wide, with a TV camera attached to a telescope ... this information would be transmitted to Earth, arriving a few minutes later, and revealing objects as small as 1,000 feet in diameter ...*

*Theoretically, this is possible, but few engineers have much faith in any space-travel undertaking in which the Earth's surface is the starting point.*

—DR. I. M. LEVITT, *A SPACE TRAVELER'S GUIDE TO MARS*, 1956

## TECHNICAL DIFFICULTIES

As much as science-fiction writers might inspire generations of scientists and engineers with their fantastic visions of the future, many times, we must admit, those visions fall a little short.

For instance, there's a classic novel by famed Canadian science-fiction writer A. E. van Vogt, titled *The Voyage of the Space Beagle*. The original H.M.S. *Beagle* was the ship on which Charles Darwin embarked on his historic voyage of science and exploration from 1831 to 1836—perhaps the original "five-year mission." Van Vogt's *Space Beagle* continued in that tradition, being the account of a spaceship's voyage of science and exploration among the stars.

There is no doubt that Van Vogt's book—a knit-together collection of four long stories, the first of which was published in 1939—is an important part of the history of science fiction. *Voyage of the Space Beagle* is one of the first depictions of a large spacefaring craft organized along more-or-less naval traditions. Clearly, this book and its tales of encounters with alien life-forms helped inspire *Star Trek* as well as the series of *Alien* movies. But it is also an important illustration of the difficulties of predicting the future.

The *Space Beagle* was an immense ship with a crew of almost one thousand, capable of traveling between *galaxies*. Yet, in that far future of technological marvels in which its voyages took place, once a *Space Beagle* scientist had dictated his report about the latest alien encountered, that recording had to be taken to the ship's steno pool to be transcribed by *typists*.

No wonder the *Space Beagle* needed so many crewmen. (And they *were* all male.)

A closer-to-home example of science-fiction writers getting close, then missing, can be found in the particulars of how Sir Arthur Clarke, author of *2001: A Space Odyssey* among many other significant novels, "invented" the communications satellite.

In 1945, Clarke published a four-page paper in the British journal *Wireless World,* titled "Extra-terrestrial Relays: Can Rocket Stations Give World-wide Radio Coverage?" In it, using charts and diagrams, he describes in detail how at least three of what he called "space-stations" placed in geosynchronous orbit around the Earth could serve as radio receivers and transmitters, so that signals could be sent up in a straight line from the Earth's surface. The signals would be retransmitted from space station to space station, then transmitted back to any other point on the surface. At the time he wrote the paper, Clarke noted that scientists did *not* have definitive proof that it was possible for radio waves to be transmitted beyond the Earth's atmosphere, but the general technique he outlined in 1945 is exactly the one used today that makes us take for granted full-color live television coverage from anywhere on Earth at any time.

On the other hand, when Clarke wrote his paper, what we today call communications satellites, he saw as crewed space stations. And the reason for that crew? To change the vacuum tubes in the radio receivers and transmitters.

In terms of the Red Planet, Clarke also came close, then missed, in his classic 1951 novel *The Sands of Mars,* one of the first to advance the idea of terraforming—that is, deliberately altering the Martian atmosphere, and thus its climate, to make it capable of supporting Earth life. Clarke himself ruefully points out that his novel contains the sentence, *There are no mountains on Mars,* in italics no less—a small detail for which he apologized when, in 1972, *Mariner 9* sent back images of Olympus Mons, the tallest volcano in the solar system. (And to prove that some things never change, one of the characters in *The Sands of Mars* is a writer from Earth who travels to a Martian colony, where he writes on a typewriter with carbon paper, then faxes his story home.)

If there's any lesson to be learned from these excursions into the future, perhaps it can best be summed up by Yoda, the Jedi master of *Star Wars:* "Always in motion the future is."

Yet there is one thing that's not in motion when it comes to the exploration of space, and that is the laws of physics—at least as we understand them today.

Science-fiction writers might postulate miraculous breakthroughs like reactionless impulse engines and antigravity drives. NASA engineers are studying more likely possibilities, including fast nuclear propulsion. But the universal laws of gravitation as described by Sir Isaac Newton will ensure that Mars will always follow its orbit just as the Earth will. And however a spacecraft travels from one planet to the other, five years from now, or a hundred, it will always require a specific expenditure of energy—from whatever source engineers provide—to make that journey.

Similarly, transmissions of data from Earth to Mars must travel at the speed of light. (Though if, someday, the unlikely becomes real, and scientists figure out how to circumvent the speed of light as the universe's absolute speed limit for the transmission of information, our *description* of the universe will need to change in a big way, and a very new set of rules will come into effect to cover the new phenomenon.)

So while the technology we use for going to Mars has been and will be continually changing and improving, the conditions to be faced on that journey will remain basically the same.

The scientists and engineers who fly spacecraft to Mars, including the Mars Pathfinder mission, proved equal to the task of balancing technology against those conditions, and in one case at least, science fiction played its part in inspiring a crucial member of the team.

## THE ART OF PRECISION

At Paramount Studios in Hollywood, home of the many *Star Trek* television series and movies, the designers in the various *Star Trek* art departments decorate their walls with hundreds of inspirational images of real spacecraft. It's very fitting that a few miles away in a cubicle at the Jet Propulsion Lab, the engineer who led the navigation team that guided the *Pathfinder* spacecraft on its 309 million miles from Earth to land within twelve miles of its target coordinates on Mars has a model of *Star Trek*'s Klingon bird-of-prey spaceship displayed on top of his computer.

The engineer in question is Pieter Kallemeyn, and this is how he describes the challenge he faced as NAV (navigation) leader for the Pathfinder mission: "There is a planned trajectory and then, of course, the real one. And they never really match up unless you're really, really good and a little lucky."

Fortunately for *Pathfinder*, the navigation team was both.

Pieter's personal trajectory to becoming a spacecraft navigator began in his childhood, when he was caught up in the adventures of both real and fictional spacecraft crews. "I've been a science-fiction fan as long as I can remember. Before that, a fan of the space program. Anything dealing with space and space exploration, I really love."

Pieter was too young to closely follow the Apollo moon missions, but he does recall following *Skylab*—America's first space station—and the first joint U.S.–Russian space mission: the *Apollo-Soyuz* flight of July 1975. (As further evidence of

the way science fiction and space exploration are often intertwined, the historic *Apollo-Soyuz* flight came about, in part, because of the 1969 science-fiction movie *Marooned*, based on the novel by Martin Caidin. In the movie, a Russian cosmonaut helps rescue American astronauts stranded in orbit. In real life, Russian space officials were amazed that Americans would make a movie in which a Russian was one of the heroes. After the movie premiered in Moscow, serious discussions began between the two countries in order to standardize equipment that would, indeed, allow astronauts and cosmonauts to come to each other's assistance in space should the need ever arise.)

At the time of *Skylab* and *Apollo-Soyuz*, Pieter never thought he would actually be involved in what he calls "the space business." But he owes the moment he realized that that might be a possibility to his math teacher at Douglas County High School in Castle Rock, Colorado.

The science-fiction movie *Marooned* featured a space station named *Iron Man 1,* based on the U.S. Air Force's Manned Orbital Laboratory (MOL) plans that eventually led to the *Skylab* space station. The movie helped inspire co-operation between the Russian and American space programs, to create compatible equipment that in the event of emergency would permit crews from one country to rescue crews from the other. COURTESY THE KOBAL COLLECTION.

"My math teacher interviewed me and said that I would be a good candidate for the Summer Institute at the University of Colorado. It was held every year for high school kids to live at the University of Colorado for a week and take classes in engineering, just to see what it's like to be an engineering student." Pieter's reaction was, "Wow, this is what I want to do."

Pieter's week at the university was even more than he hoped it might be. "We looked at wind tunnels and we talked to former astronauts who also went to the University of Colorado. Vance Brand was an alumnus and he's been on the shuttle quite a few times. He also flew on *Apollo-Soyuz*. So it was really interesting."

Before that week, Pieter already had been considering the study of aeronautical engineering, and the University of Colorado became his first choice. But when he entered UC, he had a slight change in plans. "They were just starting up this program in astrodynamics, which is the study of the dynamics of the solar system—the so-called celestial mechanics of how objects orbit the Sun and how vehicles travel between the planets. I really got into that. It was like candy to me. So I studied that, and then I got hooked up with a satellite control center that was operated by students who ran a small Earth-orbiting satellite called the Solar Mesosphere Explorer. A tiny little thing."

More significant to the future navigation leader for *Pathfinder*, that "tiny little thing" was a half-ton satellite that had been managed for NASA by JPL.

Pieter explains, "The university built the satellite's instruments, Ball Aerospace (also in Boulder, Colorado) built the rest of the spacecraft. It was launched on a Delta II rocket, which turned out to be the same booster that the *Pathfinder* and MER missions were launched on. The satellite was only supposed to last, I think, six months, and this was back in '81. But it kept going and going. So every year, the university said, 'Well, we can make it more affordable—and we can get more science out of it—if we hire students rather than professionals, because students are practically free.'

"What the University would do is interview a whole bunch of students who were interested in the project, and of those they'd take about ten every year. The students would train during the summer, so for three months we would learn everything there is to know about satellite operations and orbit prediction. We were told how to talk on the communication networks, what stations we would be in contact with. And after that, they would let us 'control' the spacecraft." Pieter quickly adds with a smile, though, "We would work with a professional flight controller, so there was someone there to be sure we couldn't actually hurt the spacecraft."

Pieter's role in the student project for the university's Laboratory for Atmospheric and Space Physics (LASP) was excellent training for his later career. "There were probably about twenty students who would work in shifts, and at every moment there would be one professional flight controller and one student command controller. I was one of the command controllers."

Pieter was chosen for that position through a long selection process. "You would first put your résumé in and from that they would pick probably twenty or thirty students they would want to interview for ten positions. They got hundreds every year because it was a big thing. The first year I applied for it, I was number eleven out of ten. They said if one of the ten chooses not to do it, you're in. And they all chose to do it. So I had to wait another year. But by the next year, when I resubmitted my application, I was a senior in college, and they wanted to pick juniors because the university wanted a two year return on their investment. They only wanted people who would be there through their junior and senior year."

Fortunately, Pieter was able to make a persuasive argument. "I really wanted that job, so I told them I would stay on through graduate school, and they said, 'Okay, you've got it. So I finished out my senior year working as a CC—that's what they called us."

Pieter continued to work with the Solar Mesosphere Explorer until 1988, when he graduated with a master's degree in aerospace engineering sciences with an emphasis in astrodynamics. "I was involved with the tricks of the trade in navigating spacecraft and attitude control, but I was really looking for a job building spacecraft."

When it comes to building spacecraft, JPL is a prime destination, so after graduation, Pieter sent in his résumé blindly, without knowing if the Lab was even hiring. But he was called for an interview and promptly offered his choice of two positions—not for building spacecraft, but for "flying" them.

"They gave me my choice of joining the multimission spacecraft team that worked on a variety of different missions, or I could work with the *Galileo* navigation team. I picked *Galileo* because I was familiar with the mission [to Jupiter]."

Pieter's precise job description was "orbit-determination analyst for the *Galileo* navigation team." What that meant was he was part of the team that figures out what the trajectory of the spacecraft actually is. And that brings us back to Pieter's description of the challenge he faced as the navigation leader on *Pathfinder*, regarding a spacecraft having a planned trajectory and a real trajectory, with both only matching up if luck is added to the equation. It turns out that's true for every spacecraft, not just those going to Mars.

Pieter describes the teamwork required to resolve those differences. "The job of the trajectory analyst is to figure out what the planned trajectory *ought* to be. The job of the orbit-determination analyst is to figure out what the actual trajectory *is*. And then we have the third part of the NAV team, who is the maneuver analyst, and he figures out how to get you from where you are to where you want to be."

It's fascinating to hear spacecraft navigation described in this way. Much more typical is the kind of announcement NASA made on June 30, 2004, when it stated that the *Cassini* probe achieved its predicted orbit around Saturn by firing its engine for a period within *one second* of the predicted time. Achieving a level of accuracy to within a second after almost seven years of traveling through space certainly gives the outside public the impression that when it comes to spacecraft navigation, everything runs with clockwork precision. To Pieter, though, spacecraft navigators are the last to see their job in this way.

It seems everything depends on how "precision" is defined.

Pieter says it's a matter of degree. "The public sees being able to land on Mars

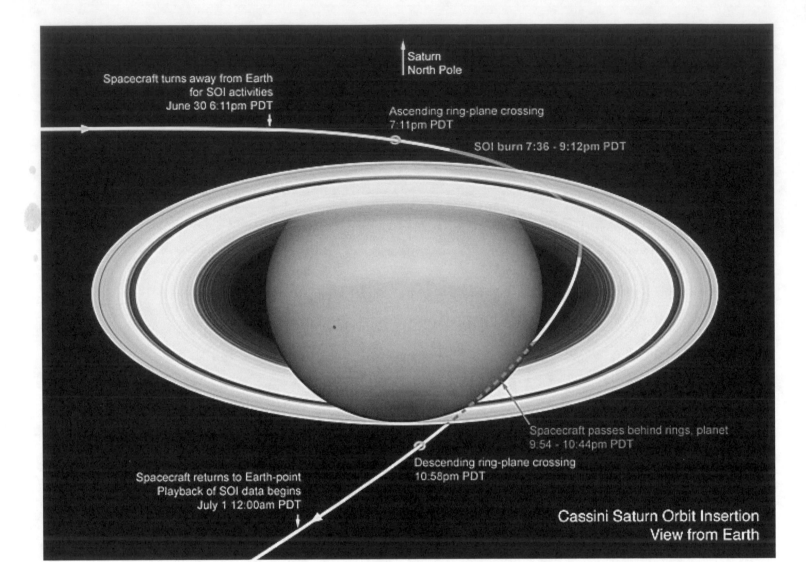

Saturn
North Pole

Spacecraft turns away from Earth
for SOI activities
June 30 6:11pm PDT

Ascending ring-plane crossing
7:11pm PDT

SOI burn 7:36 - 9:12pm PDT

Spacecraft passes behind rings, planet
9:54 - 10:44pm PDT

Descending ring-plane crossing
10:58pm PDT

Spacecraft returns to Earth-point
Playback of SOI data begins
July 1 12:00am PDT

Cassini Saturn Orbit Insertion
View from Earth

**Precise navigation permitted NASA's _Cassini_ probe to "thread the needle" of Saturn's rings, precisely traveling through a narrow gap in the rings after a voyage of more than seven years. COURTESY NASA/JPL.**

within twelve miles of a predetermined spot as a really big thing, a great achievement. And you can do that, provided you have enough data available and you have enough thrusters on board the spacecraft to make any minor corrections."

This is where real precision comes in. "We deal with forces so small that in typical everyday life you don't notice them at all. Like solar radiation pressure. If you're out in space, if you were to shine a light on something, the photons bouncing off that object have a certain momentum to them. And that will cause the object to move if no other force is acting on it. A good example is the 'radiometer' toy you've probably seen in hobby shops: vacuum jars, shaped like a lightbulb, which have four vanes inside looking like small flags, each with a white side and a black side.

"The black side absorbs the photons, while on the white side some of them bounce off. So, when the photons bounce off, they impart twice their momentum to the white flag. That makes the vanes actually spin in that direction, propelled by nothing more than solar radiation pressure."

What happens to the radiometer toy in Pieter's example also happens to multi-ton spacecraft flying through absolute vacuum at tens of thousands of miles per hour. Amazingly—at least to those not in the space business—the huge craft can have their trajectories affected by something as seemingly insubstantial as sunlight!

Pieter continues. "And that's just one of the forces that we have to account for when we do spacecraft navigation. Now, if you study the motion spacecraft long enough and you look at it with the tracking data that we have available through the DSN [Deep Space Network], you get to understand how the spacecraft is behaving, and how solar radiation pressure and other forces are affecting it.

"We know the gravity of the planets extremely well and that's actually ninety-nine percent of the job right there for getting us to Mars. It's understanding the final touches that can get us down to within that ten miles or so of the target that's the challenge. So we model the gravity. We model solar radiation pressure. And another big uncertainty is the thrusters on the spacecraft. (It was an error in the model of the thruster firings—an English-to-metric-system error in the conversion of thruster forces—that was the initial mistake that resulted in the Mars Climate Orbiter missing its arrival point and burning up in the Martian atmosphere.) Every time the *Pathfinder* spacecraft did a turn, it would fire a number of thrusters. They were supposed to be balanced, but, of course, we could see some very small unbalanced elements. So we had to model and account for every one of those. A navigator's dream job is a bowling ball with absolutely no thrusters, very low solar pressure, just moving along nice and clean."

Once again, when it comes to spacecraft design, things are not what they seem, at least not to the average person. We're all familiar with the sleek aerodynamic shapes of high-speed aircraft, necessary for traveling through the atmosphere. Yet many of us take for granted that the spindly, exposed, distinctly non-aerodynamic shapes of spacecraft designed to travel outside of the atmosphere—like the *Apollo* lunar modules and the *Cassini* Saturn orbiter—indicate that atmospheric drag isn't an issue and that the craft can therefore take on almost any shape. However, when dealing with the level of accuracy Pieter and his navigators have to achieve, the shape of non-atmospheric spacecraft *is* critical after all. "Not to a degree that others might notice," Pieter says. "But to navigators, if they're trying to reach this small spot, that's a big concern."

How big a concern? Well, for *Pathfinder*, "Just the level of uncertainty that we had in solar radiation force would have amounted to missing Mars by maybe one hundred or one thousand kilometers if we didn't have the solar pressure and space-craft shape modeled correctly." That's between 63 to 630 miles, which brings to mind the near misses of the space probes of the 1960s, and might provide at least a partial explanation for those early losses.

"I don't know what the history of those losses is," Pieter admits. "But every spacecraft that we launch, we've learned the lessons from the previous one. So though I'm not sure about the days of *Surveyor* [lunar probes], I know, certainly, by *Mariner 10* [1973–1975], that was a spacecraft where solar pressure was under-

## AMBULANCES IN SPACE

*How Doppler-Shift Tracking Works*

JPL's record in hitting its celestial targets has been extraordinary. For the Mars Exploration Rovers of 2004, the navigation team hit their entry corridor in the Martian atmosphere within less than a mile of their target after a journey of more than 300 million miles. That's like making a hole-in-one from Los Angeles to Houston, and even Tiger Woods isn't that good.

Given that there are no Global Positioning Satellites—GPS—in deep space to trilaterate (similar to triangulate) our spacecrafts' position as they travel from the Earth to distant planets, comets, and asteroids, how do JPL's navigators manage to hit their target coordinates so precisely? Amazingly, they can do it basically by starting with only three pieces of information (there are more advanced techniques in use, but those are beyond the scope of this sidebar):

1. The time it takes for a radio signal to travel from the Earth to the spacecraft.

2. The angle that the signal comes from relative to the Earth.

3. The rate at which the signal changes when the spacecraft changes its velocity.

Spacecraft navigators use these three measurements along with their knowledge of celestial mechanics and a very sophisticated piece of software called the Orbit Determination Program (ODP), developed over many years to predict where a spacecraft is and where it's going.

To use a radio signal to determine how far away something is requires only a simple calculation. Radio waves travel at the speed of light, which in space is about 186,000 miles per second (or 669,600,000 mph). So all you

Because the Moon has no atmosphere, the *Apollo* lunar module did not need to have an aerodynamic shape in order to safely transport astronauts. This lunar module trainer reflects the actual lunar module's ungainly—but functionally successful—appearance. COURTESY NASA.

stood because that one flew to Venus and Mercury where solar pressure is a pretty intense force."

For the technically minded, solar pressure varies according to the inverse-square law. For the rest of us, that means, the farther you move from the Sun, the weaker solar pressure gets. (The inverse-square law allows scientists and engineers to calculate exactly what the decrease in pressure is.)

For Pieter and other spacecraft navigators, the bottom line is, "Something like *Galileo* at Jupiter really isn't subjected to nearly the pressure you might see at Mercury. So the way we navigate is we have computers on the ground running very computational-intensive software that models all forces on the spacecraft and calculates out of that, the flight path. We store the predicted flight path, and then when we receive tracking data from the DSN [Deep Space Network], for every point of data we get, there's a specific value. We can measure the position for a spacecraft *very* accurately now."

---

One of the techniques that's used to determine the spacecraft's position is similar to the way an echo works. The ground sends a radio tone to the spacecraft, which the spacecraft sends back like an echo. Then the ground computers note the precise direction the echo came back from, and compare the time of the echo's return with the time it was transmitted. The navigation team compares that measurement with the measurement they would have received with their predicted trajectory, then subtracts the two to get what is called "a difference or residual." If that difference equals zero, that means the spacecraft is flying the predicted trajectory perfectly. However, in reality, the difference is never exactly zero. There's always a small error. But by studying the residual and the spacecraft's behavior, the navigation team can make subtle changes to the models that affect the flight path. The object is to get those residuals to be as close to zero as possible.

---

For *Pathfinder*'s flight to Mars, the correlation between the predicted trajectory and the actual trajectory was very close—"to within a fraction of a millimeter per second," Pieter says. As we've noted, though, scientists and engineers tend to be skeptics by nature, and good results are just as subject to ongoing scrutiny as bad ones. Thus, "For the first thirty days of flight, the *Pathfinder* spacecraft was monitored nearly twenty-four hours a day. And when we monitor it, we also collect the tracking data. So we had enough to keep us busy every day for the first thirty days of cruise.

"After that, we went down to only three days a week that a DSN station would look at the spacecraft, so we were only working maybe half the time. But, still, we were doing work in between the passes. Looking at the data. Assessing it. And if a [trajectory-correction] maneuver was coming up, we would plan for them. Then, in the last forty-five days, we went back to the twenty-four-hour coverage. Overall, we had four key maneuvers: one at thirty days after launch, one at sixty days after launch, one at sixty days before arrival, and the last one at ten days before arrival." And just to be absolutely certain, the Pathfinder team had also built into the mission plan the possibility of a fifth course-correction maneuver within ten *hours* of arrival.

Was it necessary?

Pieter describes the final leg of *Pathfinder*'s flight. "I remember the last two days before approach, the whole NAV team—and there're only three of us: myself,

need is a watch and the fact that the speed multiplied by the time equals the distance traveled. A signal is sent from Earth to the spacecraft where the onboard radio takes that signal and turns it around and sends it back to Earth. The time traveled (minus the few milliseconds it takes for the spacecraft's radio to receive, process, and retransmit the signal, as measured on Earth before launch), divided by 2, times the speed of light, gives you the distance. (Trust me, I'm an engineer.)

For a typical Mars distance from the Earth of about one astronomical unit (about 93 million miles) it takes that radio signal about eight minutes to travel one way. If you knew the direction you were heading when you left Earth and nothing disturbed your path, this simple calculation would tell you where you are. But it should come as no surprise that space navigation isn't quite that simple.

The other key piece of information that navigators use is the Doppler shift, named after the Austrian scientist Christian Johann Doppler, who first predicted the effect in 1842. The Doppler shift is the characteristic of waves, for example, sound waves, that we notice when either we or the source of a sound are moving relative to each other. To use the tried and true analogy of an ambulance siren—the pitch of the siren becomes shorter (that is, gets higher) as it approaches us and longer (gets lower) as it leaves us.

This same phenomenon occurs with electromagnetic waves, which include light and radio. According to Newton, and he's been proven right many times, a body in motion tends to stay in motion along a straight line unless acted upon by a force. In other words, if a force (like gravity, or solar pressure, or the spacecraft's thrusters) act on a spacecraft, that action changes the spacecraft's speed and/or direction, and we can measure this effect through the Doppler shift in the spacecraft's radio signal. We then take the information from the Doppler shift and add it to our other calculations to estimate how our speed and direction have changed.

As Mars *Pathfinder* and the MER spacecraft reached their target, the project team watched our computer monitors intently as the radio signal from the spacecraft gradually shifted to a longer and longer wavelength. That told us that Mars's gravity had hold of the spacecraft and their velocity was increasing as they began to fall faster and faster toward their destination. (This acceleration effect is most prominent in the last few hours before landing). The constantly changing Doppler shift was monitored carefully by the navigation team and used to verify that the trajectory was as they predicted.

Sometimes, however, the information provided by the Doppler shift doesn't confirm our predictions. In the early morning of September 23, 1999, I talked with one of the project leaders of the Mars Climate Orbiter mission. It was the fact that the MCO's Doppler shift was much greater than expected that told him his spacecraft was coming in too close to Mars, too low into the atmosphere, and that it was doomed.

—B.M.

Robin Vaughan, and Dave Spencer—we were awake the entire time and looking at the tracking data, just making sure that we were on course and that we didn't see the solution start migrating."

When the time came to decide whether or not to use the fifth and final course-correction opportunity—very risky because there would be no time to compensate for any errors, should any occur—an examination of the probable landing area within the 60-by-125-mile landing ellipse revealed the presence of a hill. "I remember going to Mission Director Richard Cook and asking, 'Is that going to cause us a problem if we land on that?'"

The odds of actually hitting the side of the hill were fairly small, but the potential for damage to the airbags, perhaps resulting in the loss of the mission, was significant. "So there's this small amount of worry multiplied by a large amount of worry that it could cause us a problem if we landed there."

Richard Cook called in mission scientist Matt Golombek, who examined the best photos and mapping data available for the site that he had been instrumental in choosing. (The photos were twenty-year-old *Viking* images, at one-thirtieth the resolution we have now.) "This was all in the course of just a few minutes," Pieter reminds us. "Matt had to go back to his notes, check this region, and meanwhile we're waiting. The clock is ticking. Our uplink window for TCM 5 [the final trajectory-correction maneuver] is coming up.

"Finally, Matt comes in and he says, 'You know, it's not as steep as you might think. It's actually pretty shallow.' So we said, 'Okay, we go without TCM 5.' We actually had two windows for TCM 5. At the first window we said, 'There's no reason we should do it. Let's wait until the next window and see if the situation changes.' It didn't change. So we never did it. But we were ready."

After seven months of cruise, five minutes of terror, and a near-flawless Entry, Descent, and Landing sequence, Pieter Kallemeyn and his navigation team's precise targeting brought the *Pathfinder* down on Mars well within its targeted landing ellipse, only about seventeen miles southwest of its targeted landing site. The date was July 4, 1997. The time in Los Angeles, 10:07:25 A.M.

For the first time in twenty-one years, a spacecraft had successfully landed on Mars, using a unique, low-cost landing system that had never been tried before. As Jennifer Harris Trosper said, the whole Pathfinder team was in a state of "unbelief."

Guided by a fan of science fiction, *Pathfinder*'s flight to Mars had been a complete success. Now it was time for the next stage of the mission—ground operations.

And whether or not those operations would be equally successful would depend, in part, on how well another crucial member of the team had done his job tormenting the other members over the past year and a half.

His name is Dave Gruel, and though his original job function was officially described as "test-bed mechanical engineer" and he later became a "fault-protection engineer" and ultimately flight director, he is best known to everyone on Pathfinder as "the Gremlin."

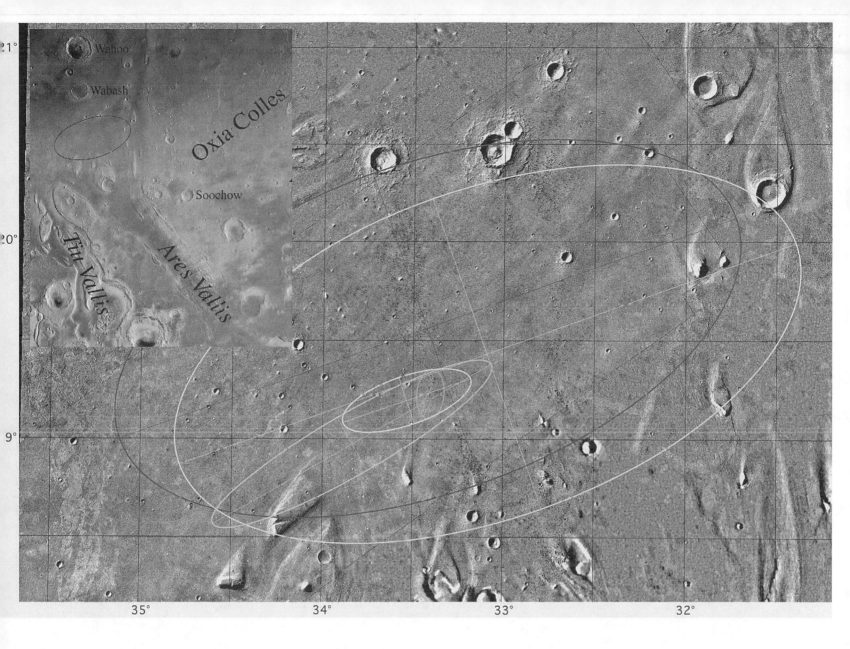

The stretched-out ellipse shows the projected landing area for Mars *Pathfinder*. The smallest ellipse shows the actual landing site, about 17 miles from dead center. **COURTESY NASA/JPL.**

## SANDBOX ADVENTURES

Dave Gruel's personal journey to Mars is as unusual as his job.

To begin with, after high school, his education abruptly stopped.

As Dave puts it, unlike most of his colleagues at JPL, in high school he had no specialized interest in space exploration, "I had a specialized interest in partying."

Refreshingly blunt, Dave expands on his high school ways. "My goal was to find the easiest way out. I didn't want to do any homework. Didn't want to study. Didn't want to do anything. My senior year of high school, I wanted to go out and have fun with my friends. That's all. Everyone else was taking their college entrance exams, writing these huge long essays about who was the one person in

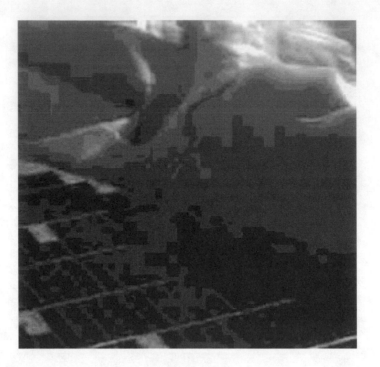

July 4, 1997. The historic first image returned from the surface of Mars by the *Pathfinder* lander—the first from the surface of Mars in almost twenty years. COURTESY NASA/JPL.

history they wanted to have dinner with and why. But I thought, 'That's too much work.'"

And people think that engineers and scientists are different from ordinary people.

Much as Brian Muirhead embarked on a series of travels and adventures before deciding what he wanted to do with the rest of his life, Dave's personal trajectory had no target after high school as well. He did end up in a junior college, but only after "Taking many, many wrong detours. I actually enlisted in the Marines for a while. I worked third shift, stocking shelves in a grocery store. But one morning I just woke up and said, 'I need something more out of life.'"

Dave's desire for something more took him to White Oak Community College in Massachusetts—one of the few options available to him because of the grades he had neglected in high school. It was while filling out his application form that his future career got an assist from that old standby—random chance.

"I didn't know what I wanted to do. I just knew I needed to do something different. So, I was filling out the college application. It's a one-page thing. I was doing really well until I came to the line that said, What's your intended major? I had no idea.

"I was looking for a challenge, so I asked the person behind the counter, 'What's the most challenging thing you've got?' She said that she sucked in math, so she imagined engineering would be pretty difficult. And I said, 'Well, my history sure backs that up, so I'll try engineering.'"

Little did Dave know at the time, but a star had been born.

Asked if he remembers any instructor in particular who helped him direct his interests and refine his goals during his two years at community college, Dave answers that all his instructors were exceptional. "Going to community college is probably the greatest thing out there that most people don't know about. It's cheap. Small classes. It's one-on-one attention. It's really a good deal."

With his self-directed goals and the support of a dedicated faculty, Dave's own talents came to the fore. Receiving an A in calculus was a powerful confidence builder. It was time to set his sights on even higher goals.

"Community college is just a two-year deal, so I knew I'd have to transfer somewhere if I wanted to finish up my degree. It was just a question of where. Now, Whole Oak usually funnels people out to the University of Massachusetts. But I said I wanted to take a step better, I wanted to try something bigger. So I actually tried to get into MIT." MIT's response? "They slammed the door in my face and said, 'Yeah, right.'"

Being turned down by MIT wasn't a complete surprise. No one had ever gone from Whole Oak to MIT before. But then again, no one from Whole Oak had ever gone to Rensselaer Polytechnic Institute, either, and that's exactly what Dave Gruel did next.

Rensselaer, located in Troy, New York, is a prestigious technological school that's been operating for more than 175 years. In fact, it's the first degree-granting technological university in the English speaking world. But Dave's initial reaction to its hallowed halls was not what one would expect, especially after the effort he had expended to get there.

"I hated it." Why? "It's a geek school."

But with the benefits of hindsight and a successful career at JPL, Dave looks at RPI a different way today. "I have to admit I was in a field full of people stereotyped as geeks. You know, walking around with pocket protectors and things like that. And coming from my high school background, some habits were still hard to break. I still enjoyed going out and having fun with my friends, and a lot of them had gone down to college in Boston. So, in those years I took off between high school and going back to college, I was down in Boston probably twice a month, going out, having fun, enjoying myself.

"But then I went to RPI hoping it would be the same thing, and it wasn't. I mean, professors purposely gave you these fifteen-page lab reports on Friday, which you had to have done on Monday. The workload just blew me away."

But Dave stuck it out, and found an area of engineering he enjoyed. "I knew I was going to do mechanical engineering. I sucked at electronics. I was horrible. I hated it. I didn't understand any of that electronics stuff." That admission brings on laughter today, because years later, employed at JPL, "That's all I do now. Circuits and things like that. But back then I was just horrible at electronics."

Oddly enough, mechanical engineering had been a fallback position for Dave. His original choice had been aerospace engineering, but a faculty advisor at Rensselaer had steered him away from the field, suggesting that aerospace might be too specialized. There could be more job opportunities with a general mechanical

background. Indeed, it was because of Dave's degree in mechanical engineering that he was hired by JPL.

Dave was considered one of the "babies" of the Pathfinder team, because *Pathfinder* was his first job at the Lab. His original assignment was to work in what is called a "test bed"—the elaborate and often messy-looking installation in which electronic equipment and software are checked out and run through their paces, first as separate components, and eventually as a fully assembled subsystem. Right next to the clean room electronics area was one of the more interesting test-bed installations: "the sandbox"—a room filled with red sand and rocks that served as Martian terrain for test versions of the *Pathfinder* lander and *Sojourner* rover.

Dave's job was to set up the test facilities and handle the installation of hardware as it came in. But like Jennifer Harris Trosper, Dave Gruel found his job expanding as he took on testing assignments and responsibilities that attracted his interest and needed to be done. The team was shorthanded for someone to debug software, and under Pathfinder's innovative management, Dave offered to give it a try and proved invaluable.

Then the team lost a critical member—the fault-protection engineer—to another job.

As so many engineers have said, testing a spacecraft is absolutely crucial; second chances in space are few and far between. Some tests are relatively straightforward. For example, from decades of flight experience, engineers know the temperature extremes a spacecraft will be subjected to on its journey to Mars, as well as the stresses experienced at launch. Thus, assembled spacecraft are put through what's called "shake and bake" testing.

First, the spacecraft is bolted to a vibration table that reproduces the shaking it will experience at launch. It will then be subjected to the acoustic noise produced by its booster rocket—142 decibels. (A thousand times louder than front-row seats at a rock concert, which can cause hearing damage, just ask Pete Townshend.) Finally, the spacecraft is shocked with thousands of g's by firing the pyrotechnic devices that hold the spacecraft together on the way to Mars, but must be released during entry, descent, and landing.

Next the craft will be placed in JPL's Space Simulator Facility for solar, thermal, and vacuum testing. The enormous vacuum chamber re-creates the conditions the spacecraft will experience in space, exposing it to temperatures ranging from −330 to +160 degrees Fahrenheit, along with intense light that duplicates the unfiltered sunlight of space—the facility is capable of producing as much as ten times the intensity of a sunny day in California.

But other tests are less clear-cut. What if part of the deflated airbags snags a lander petal? What happens if the panoramic camera turns a certain number of degrees in one direction, while the radio's antenna needs to turn a certain other number of degrees in another direction, while a stray cosmic ray forces a computer reset? The more complex the hardware, the more possible "states" each part of it can be in, and the greater the chance that some of those states will interfere with each other in unanticipated and potentially disastrous ways.

(OPPOSITE PAGE) The aeroshell containing *Spirit*—aka Mars Exploration Rover A—in JPL's Space Simulator Facility, where it will be subjected to the vacuum, the intense solar radiation, and the chilling cold of space—all at the same time. COURTESY NASA/JPL.

## "TEST, TEST, TEST!"

*How to Buy a $50 Modem for Only $500,000*

"Single-string" spacecraft are by definition one failure away from disaster. In other words, the failure of just a single resistor, capacitor, integrated circuit, or mechanism can cost you the mission. Given there are thousands to millions of such potential failures in a spacecraft, how do you protect against their failure when there's not enough time, money, and launch mass to add duplicate backups? The key is in the phrase Pathfinder Project Manager Tony Spear emphatically stressed: "Test, test, test!"

Much more than "Faster, Better, Cheaper," or "Take risk, don't fail," "Test, test, test" was the mantra of the *Pathfinder* and Mars Exploration Rovers teams. In fact, because the systems were single string, the natural tendency (consciously or unconsciously) to rely on a backup system was not present, and people paid even greater attention to being sure all the right testing was planned and conducted. Rather than limit testing due to increased cost concerns, testing was increased to control risk.

On missions where funds are limited and cost control is paramount, creative solutions to difficult problems are worth their weight in titanium. But creative solutions must be thoroughly proven before they see the vacuum of space.

A particularly interesting story is of the wireless communication system between the *Sojourner* rover and the *Pathfinder* lander.

Early in *Pathfinder*'s development phase a battle was fought over whether the rover should be tethered or free-ranging. In the early '90s, wireless technology was nowhere as sophisticated as it is today. In fact, the difference between then and now can be thought of as the difference between your cordless phone and your cell phone.

A tethered rover would be permanently connected to the lander by a physical cable that carried power and communications. In that configuration, the rover would not be weighed down by heavy batteries, which could be left on the lander, and communications would not be affected by interference or improperly tuned antennas. However, a tethered rover would have considerable difficulty in ever backing up, and the long cable dragging behind it would be in constant peril of snagging on a Martian rock.

The free-ranging rover concept eventually carried the day, requiring a pair of radio modems—one for the lander and one for the rover. Unfortunately, the technology for a very low-mass and low-power wireless radio modem that was "space qualified" did not exist.

The rover team went to private industry to see what was available commercially and then set out to see if any of those units could survive the vacuum (and near vacuum), heat, cold, and radiation of space and on the surface of Mars. This involved buying hundreds of commercial Motorola RNet 9600 radio modems for $50 each, then spending thousands of work hours subjecting them to all the physical extremes they'd encounter in space and on Mars in order to test and select the best ones. In effect, the cost of the modem pair that flew on *Pathfinder* was $100 to buy and about $500,000 to fly! But even at $500,000 it was a bargain in time and money compared to the effort and expense that would have been required to create a whole new technology from scratch.

A Mars *Pathfinder* test in the "sandbox." COURTESY NASA/JPL.

The job of the fault-protection engineer is to predict the unpredictable, determine how and why and under what conditions the spacecraft might get into trouble, and then devise strategies for either avoiding or reacting to those problems while keeping the spacecraft safe. In deep space, the quality of the fault-protection engineer's work is the mission's primary safety net. He or she must design the net and prove it works under every conceivable circumstance. That means extensive testing. Most of the time it's testing hardware and software—mostly software. But sometimes you need to test the people, too.

When it came to figuring out how good things might go bad, Dave Gruel was the right guy in the right place at the right time.

In Brian Muirhead's words, Dave possesses "a malevolent genius for devising devious 'bet you can't guess what's wrong now' scenarios" for the engineers who would be operating the spacecraft, rover, and lander.

Dave's insightful though sometimes warped approach to the operational readiness tests earned him his unofficial job title of "the Gremlin," a name passed down through the engineering community since it was popularized by World War II air crews, referring to the imaginary creature who creates problems in otherwise reliable hardware.

Dave's approach to *Pathfinder* tests has also been described as "near pathological," leaving the spacecraft operators fuming, even as they learned the limitations and capabilities of their hardware and software, and of themselves.

Brian Muirhead recounts one particularly maddening test. "Before *Sojourner* could begin her journey off the lander, the rover team first had to satisfy themselves the lander had come to rest in a safe condition, so the two ramps for the rover could be safely deployed. The view from the stereo camera on the mockup lander in the sandbox showed a three-dimensional view of the ramps that indicated they were at a steep angle, but within acceptable limits. So the rover drivers were ready to command the vehicle to drive off."

At the time, Brian was well aware of how Dave operated. The test scenario seemed a little too easy.

"Richard Cook and I suggested to the rover team that they move more cautiously, and think the situation through, to make sure they were making the right decision." But as Brian recalls, those words of caution annoyed the confident rover team. They had analyzed the situation and were convinced they were good to go.

As Brian says, "They were convinced but entirely wrong. The Gremlin Dave Gruel had created a very dangerous situation for them—tilting the lander at a steep and very *unsafe* angle. As it turned out, the software that was designed to help determine the steepness of the ramps had an error in it and was giving the rover team incorrect answers. Running a couple of calibration checks would have shown that something wasn't right."

The result was a failed rover deployment but a very successful test. Now, thanks to *Pathfinder*'s "Gremlin," those crucial calibration checks would be performed on

## ONE STRIKE AND YOU'RE OUT

*The Power of Ownership*

I believe that one of the most important concepts in any job is one's sense of "ownership." Ownership is not about your financial stake, but about commitment and integrity. The commitment to get the job done and the integrity to be sure it's done right. In the most general sense, it's one's personal sense of the commitment and integrity required by their role—no matter how large or small—to accomplish a project, and do so at the highest level of quality possible. This concept and the reality of "ownership" has been essential to JPL's and, in fact, the entire aerospace industry's way to doing business from the beginning.

As Tom Young (a member of the Viking team and former senior executive of Martin Marietta Corporation) has so appropriately noted, "We're in a one-strike-and-you're-out business." A single solder joint poorly made, a bolt not properly tightened, an error in one line of software—any one of those minor errors can cost us a billion-dollar mission. And the interesting thing about it is that every member of the team knows it. Quality control is essential; we review everything thoroughly; we check and double-check and test to exhaustion; but in the end it's each individual's personal and strongly held sense of ownership of his or her part in a complex mission that ultimately makes us successful. Everyone knows, but sometimes management forgets that quality and performance are designed and built in, not "inspected in." Ownership shows up when an engineer, technician, or a manager refuses to say "That's not my job" or "That's close enough." I saw ownership at work when, just weeks before *Pathfinder's*

Mars, no matter how safe the ramps appeared to be. Dave had helped find and eliminate one possible disaster. There were only a few thousand more to go.

Of course, not every test of the *Pathfinder* hardware and software was related to demonstrating the technology could survive disaster. Thanks to Matt Golombek, *Pathfinder* also had a science mission to perform, and all those procedures had to be tested as well.

Jennifer Harris Trosper recalls that during one test, the first photographs "returned" from the lander revealed a large philodendron plant growing in the Martian sandbox. Naturally, the plant became a science target, and the Pathfinder imaging team determined that the healthy green plant Dave had planted for them was actually red.

A life-or-death disaster was not averted by that test, but many issues related to color filters and image processing were cleared up as a result, helping ensure that *Pathfinder*'s real images from the surface of Mars would be of the highest color fidelity possible.

Under "the Gremlin," the Pathfinder team completed only one Operational Readiness Test without significant problems (though important lessons were learned from each one), and that happened to be an ORT the media had been invited to watch.

Certainly because of all the different forms of failure Dave made the team face on Earth, failures on Mars were avoided. Dave Gruel had successfully faced the challenge he set for himself so many years ago while filling out his college application form and he became a key contributor to the Pathfinder team's successfully facing the challenge of going to Mars.

## GETTING WITH THE PROGRAM

Hundreds of millions of miles from Earth, spacecraft pretty much have to think for themselves. That means good hardware is only one of the components required for success. Equally necessary is software that is designed to respond to changing conditions without requiring immediate input from ground controllers on Earth. And that's why spacecraft engineers come in three flavors—hardware, software, *and* systems. (Systems engineers are the guys—and gals—who can put the hardware and the software together and make them sing.)

Glenn Reeves is the software architect for the Mars Exploration Rovers, and like many of the key people on the MER mission, he first went to Mars with *Pathfinder*, as leader of the software development team.

We've met some JPL scientists and engineers who never imagined they'd work in space exploration, and others who always knew that was their goal. Yet almost all of them followed a winding path to JPL. Glenn Reeves is the notable exception. The path he followed was the most direct and straightforward of all.

Glenn grew up in South Pasadena, and he clearly recalls that when he was ten or eleven, "My mom used to tell me when we'd drive by this place, 'Glenn, that's the Jet Propulsion Laboratory. I think you should work there one day.'"

To make a long story extremely short, he did.

All right, we won't make it *that* short.

Glenn's not really sure what prompted his mother's advice when he was at such a young age. He wasn't taking apart kitchen appliances like Tom Rivellini or building rockets like Wes Huntress. He remembers having a wonderful space-themed birthday party when he was seven or eight, with stars and planets cut out of cardboard and hanging from the ceiling. But other than that, whatever was pointing him toward a career at JPL was something only a mother could sense. And, as mothers often are, she was dead-on.

Glenn shakes his head as he describes his personal trajectory. "To be honest with you, I believed my mom when she said, 'I think you should work there someday.'" As a matter of fact, Glenn started working at the Lab while he was still attending school. He calls that first job "a fluke."

"I was going to school out at Cal Poly [California State Polytechnic University], Pomona, and I had become sort of interested in computers. I was taking a class in assembly language, it was starting to get interesting, and my instructor was a gentleman from JPL named Dr. John Ward.

"Turns out he lives in South Pasadena, too. And one day, as I was heading toward the freeway, I happened to drive by the street where his house is. He was at the bus stop waiting for the bus because his car was broken. I thought, 'Heck, there's my advisor. I should have stopped and picked him up, but no . . . I made him take the bus.'"

Glenn laughs at the memory, then adds, "But I went and talked to him about it later, and he and I became friends. And eventually he said, 'Would you like to come work up at the Lab?' And I said, 'Yeah, I'd love to.'"

Up to the moment Glenn arrived at the Lab to hand in his application form, despite living in such close proximity, he had never once visited JPL, either on a school trip or to take a public tour. However, his first visit was productive, because, "I ended up with a job called Junior Programming Aide for $7.25 an hour, and I worked here part-time for two years."

Because Glenn was still in school, he was much younger than most other JPL employees, and that took its toll. "For the first two years I was here, I didn't even know that there were two cafeterias on the facility. [There are three now.] I didn't even know there was an East parking lot. I was so isolated. I came in early. I walked in from the lot. I went into the Lab. I worked with two or three guys in an isolated environment. I went to school. I went home and I did homework. And that was it. That's all I did for about two years."

For that two-year period, Glenn's job was working on what he calls a "little tiny project related to fault-tolerant computers called the self-checking computer module. I did a bunch of programming for that. Learned a tremendous amount from many of the people I still work with here."

With two years' experience at the Lab, it was no surprise that Glenn was offered a full-time position as soon as he had graduated. "This little self-checking computer module project had basically run its course and was closed down. Then I was offered a job on the *Magellan* spacecraft and mission development." (*Magellan* was the

landing, Attitude Control Lead Engineer Miguel San Martin was reviewing the parachute deployment algorithm and saw that there was a possibility, albeit remote, that under some circumstances it might not work as needed. Rather than just say, due to the time crunch, "It's close enough," we gave Miguel the three days he asked for to develop a new design, test it, and convince himself and us that he had something better and that it was necessary to make the change. He did it, we reviewed it, and we made the change.

Another example of ownership is found in the story of the technician on the Magellan mission to Venus who, after having done the final installation of the device that would ignite the solid rocket needed to place the spacecraft into orbit at Venus, had second thoughts on the plane ride back home. Rather than just put it out of his mind—Who would ever know?— he called back to the Kennedy Space Center and told people there that he thought he'd made a mistake and asked them to check the wiring connection again. Upon reinspection, it was discovered that the connection *had* been wrong and that the rocket would have never fired. But because of the technician's call, the connection was fixed on the spot and worked perfectly at Venus.

There can be times when the ownership spirit can become weakened or might even be absent. This is the time for management to be afraid, be very afraid. Lack of ownership shows up when one does not have a sense of control, real or perceived, over the job he or she is supposed to be responsible for. It shows up when people feel they're not being heard. It can show up when others aren't carrying their load and management lets them get away with it. In the aerospace business, this can occur when we are handed an over-constrained problem (which most missions are) and not given the freedom to be creative about solving it. JPL and the aerospace industry in general are not afraid of "impossible missions," in fact we thrive on them. But we succeed only when people feel ownership and are empowered to find innovative solutions to the complex technical problems and programmatic challenges inherent in the business.

Every mission to Mars has had a unique set of challenges, and whether successful or not, they all have shared an extraordinary level of individual commitment by every frontline member of the team. But after all the analyses are done, the tests completed, the paperwork signed off, individual people's ownership of their jobs—their commitment to doing their part to perfection—is still the best insurance of mission success.

—B.M.

wonderfully successful mission of radar-mapping Venus. It was the success of that mission that prompted Wes Huntress to assign one of its project managers, Tony Spear, the initial study for sending low-cast landers to Mars—the mission from which Pathfinder would be born.)

Glenn was excited to have a chance to work on an actual spacecraft, but then realized his job was to work on the *test equipment* for the spacecraft. This is the complex set of hardware and software used to simulate the operating environment in space and for testing the spacecraft electronics and software before and after launch. Close, but not quite the same as the real thing. Still, it was all part of his JPL learning curve. "I think that's at about the time when I first started to become a little bit aware of spacecraft and mission development and things like that. But I was pretty naïve. I had only worked with fairly low-level software."

But Glenn's days of isolation were over. "The self-checking computer module crew I worked with was about four or five people, but for the test equipment for this *Magellan* spacecraft, there were probably more like eighteen or twenty people. So I got a little more exposure to the Lab."

After his work on *Magellan*, in 1986 Glenn moved on to work on ground support software for the *Cassini* mission to Saturn. Then he undertook a career move not too common among JPL engineers. "I left in 1989 to go work out in the real world."

His motivation was very practical, and Glenn laughs as he recalls his reasoning. "I had been told, in general, that if you stayed here, the wage increments weren't quite as nice as you might like. But if you left, and stayed away for a long enough time, they'd 'reprice' you at a new level in accordance with the aerospace industry when you came back in."

But JPL had a hold on Glenn. "Even when I left, I didn't quite feel right about it." In fact, his new job at a nearby company required him to drive past the Lab every day. "And every time I'd drive by this place, I'd think how much I missed it. It was my home."

JPL missed Glenn, too. It took him almost eight months to leave, because as soon as he went to work for the new company, he was assigned as an outside contractor to work on some test equipment for JPL's Mars Observer Camera. (The original MOC was lost when the Mars Observer failed on its approach to Mars in 1993. But the flight-spare MOC was launched on the successful Mars Global Surveyor orbiter and has been providing a nonstop stream of high-quality images of Mars since 1999.)

That small assignment took eight months to complete. Then Glenn truly was cut off from the Lab and missing it badly. So badly that within a year and half, he was back as a contractor on the *Cassini* project, and within two years was a full-time employee again.

And, yes, what he had been told about pay increments worked out as planned. "The original premise of leaving, coming back, and getting a much greater increase than you would have gotten if you'd stayed for the duration, turns out to be quite true," he says. (This practice is no longer common.)

But as nice as it was to have a raise in pay, there was a much better benefit to coming back to JPL. "I was home. And I haven't gone anywhere since."

Happily back at work at the Lab, Glenn was nearing the end of his work on the ground-test equipment development and told his supervisor that he'd be interested in getting involved with actual flight software. That's when his supervisor said, "There's this MESUR Pathfinder thing starting up. Would you be interested in that?" As Glenn recalls: "I thought about it for three nanoseconds, and I said, 'Yeah, I really would be interested.'"

Glenn's interest in what would become the Mars Pathfinder mission was so strong, he began working on it before even its development phase was officially approved. His first contribution was to work with his good friend Rob Manning to define the requirements of the flight computer. (Anyone who remembers the happy fellow with the full beard reporting the performance of the spacecraft during EDL, then grinning ear-to-ear after the safe landings of the Mars Exploration Rovers *Spirit* and *Opportunity* knows Rob. At the time, he was the MER mission's Entry, Descent, and Landing systems manager, but he had already made his first trip to Mars as part of the Pathfinder team.)

During one of Glenn's meetings in Rob's office, Rob got a call from Brian Muirhead, who was trying to recruit Rob to be chief engineer for *Pathfinder*'s flight system.

Glenn remembers whispering loudly to Rob: "Take the job!" (which he eventually did).

At the time Glenn started with the MESUR Pathfinder program, like many of the others drawn to the project, he knew he was taking a gamble on his future. The *Cassini* mission was JPL's chief focus, employing the lion's share of the Lab's flight project workforce. *Pathfinder*, as a proposed faster-better-cheaper mission, was a true underdog, with no guarantee that it would actually be approved, let alone succeed. But that's part of the reason why the people who jumped aboard were so drawn to it.

"It was an exciting time," Glenn remembers. "It was an exciting mission to do. Even though it was all under the shadow of *Cassini*, and we were just a small little thing—almost a gnat or a wart at the beginning."

So what was the appeal of working on a wart of a mission for the Pathfinder team?

Glenn's answer: "I can only speak personally, but Pathfinder gave me an opportunity to do something new, something different, more responsibility. All of those things come into play. It's not about money, or anything like that. It's . . . about professional challenge and personal development."

Glenn started on Pathfinder by working on the Attitude Information Management subsystem, otherwise known as AIM, which is the same subsystem Jennifer Harris Trosper would eventually be brought in to test.

For AIM, Glenn's job was to develop the software that would operate on the subsystem's hardware, but since the only other computer was in *Sojourner* (which was

"asleep" for most of the trip to Mars), this would be the system that would fly and operate the entire mission.

What made this job particularly new was that this would be the first time JPL would design and write software for a "modern" computer. The Pathfinder team had chosen to stake their mission on a new computer, one that hadn't even been developed yet, but one that would allow the software team to use a modern programming language (C) and modern design and test tools. The new computer was based on the IBM R6000 processor that had been redesigned to be radiation tolerant with funding by the Department of Defense—it was called the RAD6000. This processor had to be integrated with all the other elements (memory, input/output, power, etc.) that were needed to make a computer, then tested to prove it could survive the rigors of space travel. So, effectively, Pathfinder was starting from scratch for both the hardware and software, and both were needed in eighteen months from the start of the project.

The first step in that software design process was for Glenn to present an "architectural" view of what the flight software would look like—a rough sketch to show the modules that would be required. "I put together a design for the flight software that I drew from the experience I had working on the test side of *Cassini*. And for what I was missing, I basically made it up." Glenn laughs at that admission and keeps it up as he continues, tongue in cheek. "But that was my job." It turns out that Glenn's test equipment experience was actually a huge advantage, since on the ground side he'd not been constrained the same way that flight software engineers usually are, because he'd been working with modern hardware and software for years.

# TOYING WITH THE RED PLANET

*Hot Wheels on Mars*

Since the days of the Flash Gordon and Buck Rogers movie serials, spacecraft have been a regular fixture in the toy business. But for the most part, those spacecraft have ranged from fanciful flying saucers to fictional starships. Except for plastic models of actual flight hardware designs manufactured for a small group of hobbyists, most mass-market toy spacecraft had, at best, a glancing relationship with reality.

And then came Mattel, manufacturer of the Hot Wheels series of toy cars and trucks, with more than a billion individual vehicles sold to kids and collectors over the past thirty years.

Always interested in adapting the latest vehicles into best-selling Hot Wheels toys, Mattel made a connection with Caltech and the Jet Propulsion Laboratory, which designed and built the *Pathfinder* spacecraft, as well as the first vehicle to drive on Mars: *Sojourner*.

JPL already had a Technology Affiliates Program which adapts the Lab's various engineering and software developments to the needs of private industry, but this marked the first time the lab had ever worked with a toy company.

As things turned out, it was a match quite literally made in the heavens.

Mattel's Hot Wheels Action Pack featuring the *Sojourner* Mars Rover—along with smaller reproductions of the *Pathfinder* lander and the entire spacecraft in its aeroshell—was an instant success, and for the first few months was almost impossible to find on store shelves. Proving that the public's fascination with Mars and the Pathfinder mission was not to be shortlived, four months after the rover landed on the Red Planet, Mattel had to double its production of the toy to keep up with demand.

Caltech, JPL's parent organization, benefited from the toy's success not only financially but also in terms of publicity. The Lab received a royalty from each unit sold and was able to use that money to expand its outreach program, providing materials and assistance to educators and the media to help promote space exploration. But perhaps more important, millions of people were now able to have a little piece of the space program in their homes, with the JPL logo attached to the rover and to the *Pathfinder* lander, thus increasing their awareness of the Lab and its leading position in the exploration of space.

Mattel, of course, benefited by having one of the top-selling toys of 1997 and immediately extended the line by creating an Action Pack based on the Galileo mission to Jupiter. The Hot Wheels Jupiter/Europa Encounter set included reproductions of the *Galileo* spacecraft, its Jupiter descent probe, and a giant antenna dish from the Deep Space Network used to communicate with NASA's spacecraft. And JPL has since become even more involved in the toy business by licensing its logo not only for models and Lego construction sets based on the 2004 Mars Exploration Rovers, but to a line of Mars Explorer toys, based not on hardware being built today, but on designs for the future, when humans will accompany machines to Mars.

The *Pathfinder* rover's appeal extended past Hot Wheels. Other toy companies brought out different versions of the little machine, ranging from elaborate radio-controlled models to inexpensive, windup playsets. Demonstrating the power that the idea of going to Mars has among people of all ages around the world, the rover had gone from being just another piece of flight hardware to becoming an icon of the space program.

Perhaps when the first people to walk on Mars leave their footprints in the red soil of that world tomorrow, they will remember playing with, and being inspired by, the Mars toys of today.

## A MASSIVE CHALLENGE

*The Importance of Accounting for Every Ounce*

Mass. The most basic property of physical stuff. Mass is what we measure, and agonize over, when we stand on a scale. Mass is what makes things hard to lift, especially for rockets. Most "rocket science" is just about finding the most efficient way to lift "mass" off the Earth into space. The more massive your spacecraft, the bigger (and more expensive) the rocket you need to get you off this rock. Mass (also called inertia) is what makes things hard to speed up and slow down. Think of the difference between catching a watermelon and a lemon. The more massive a thing is, the harder it is to throw and to slow down and catch. Feel free to try this experiment at home, preferably outside.

The management of mass is a foreign concept. In our day-to-day experience, many of us worry about managing mass only when we're looking to buy a swimsuit or see an ad for Jenny Craig. But for space missions, it is one of the most challenging parts of the job. Spacecraft builders generally start with a mass allocation, derived from the mission requirements (where you're going and when) and the launch vehicle options available to you. Once you pick a launch vehicle (generally the biggest rocket booster you can afford but not as big as you'd like) and select a launch and arrival date, your launch mass allocation is fixed. And, of course, this number is picked well before you have much of a spacecraft design. So this is like the marketing department telling the engineers what the curb weight of a new SUV will be (along with all the sexy features they want) before the engineers have any idea what an SUV looks like.

So how do spacecraft builders stay within their mass allocation (forget the mental images of refrigerators full of Weight Watchers and Lean Cuisine)? We start with our best idea of what our payload will be (science instruments, cameras, et cetera) and what subsystems (electronics, propulsion, sensors, radios, et cetera) we will need, and then wrap them in structure and figure out how to keep them from getting too hot or too cold. Pretty straightforward but . . .

From this first guess of a concept, and our experience, we develop our best estimate of what the spacecraft will weigh at the time of launch, years away. We call this number our current best estimate, CBE, and the difference between the CBE and the allocation is the "margin." The margin must cover all the changes and growth of the vehicle as it becomes better defined and as it is finally built and weighed. Like our own personal mass (weight) as we grow older, it only seems to go one way: up. We have some historical guidelines for the size of the margin we should have at any one time, and it ranges from 10 to 50 percent depending on the heritage and complexity of the system. But invariably, like money and schedule, we never have quite enough mass margin and we must manage this margin with great care.

Failure to manage the mass can cost you

---

Glenn prepared that first description of the flight software architecture in two and a half days. "It was a very short period of time, but I had been thinking about it for some time. I just had to draw it and look at it for a while."

Impressively, the essential structure of that first presentation—"made up" or not—changed very little over the course of the mission's development, eventually leading to 155,000 lines of computer code (although Glenn's first estimate was 30,000 lines). Which raises the question: Was the software so successful because Glenn's design was so good, or because his team was?

Glenn modestly suggests the answer. "I think that in general there was nothing obviously wrong with the design. I had based it on my experience of what had worked well on the ground, so why shouldn't it work in space?"

The key element, besides the RAD6000 processor, was selecting a new operating system—a real-time system called VxWorks from Wind River Systems. This operating system (which had never been used in a deep-space mission) allowed the software team to take full advantage of the horsepower of the computer and implement an object-oriented, multitask design. For the non-technically-minded among us,

the launch (because the rocket can't lift you)—or a lot of money. We always try to start with a healthy mass allocation, but we also always seem to run through it and find ourselves sooner or later needing to put the spacecraft on a diet. Diets can be expensive things. On the *Galileo* mission to Jupiter the project got into such big-mass trouble fairly late in the design process, that they were willing to pay as much as $50,000 to save a little over a pound. We're talking a very high-end fat farm.

On the Mars Exploration Rover mission they started out with a relatively low-mass margin that was justified by the belief that they were going to build a copy of Mars *Pathfinder*—whose mass was obviously very well known. However, as the team was forced to deviate from the *Pathfinder* design, the mass and its uncertainty grew. They were forced to spend money to optimize designs and use exotic materials (including a graphite composite lander structure) to keep the mass margins "healthy." At every step in the development process the mass margin is reported and scrutinized, and when it's below the desired level, the project must put a lot of engineering and management attention on it. Inevitably most projects must designate a person to be the mass czar (actually more like a Cossack) to manage this limited resource. On Mars *Pathfinder* we didn't have the money for expensive diet plans so we had to watch our weight from the beginning. I appointed myself as mass czar in order to be directly on top of the mass issues and to be able to direct changes early enough in the design so we wouldn't end up with a *Galileo*-like problem.

It turns out that the real mass limit for Mars lander missions isn't the launch mass—it's the mass you can enter into the atmosphere. The magic number for entry is called the ballistic coefficient, which we call beta. (If you must know, beta is the mass divided by the cross-sectional area of the spacecraft times the drag coefficient.) The higher the beta, the harder it is to slow down the spacecraft, and the higher the heating rate will be during entry. Our limits on a safe beta were pretty low, but Richard Cook and his navigators did a great job, continuously refining their analyses for how precisely we could target Mars and, therefore, how much mass we could safely enter.

The moment of truth comes as the hardware comes in, and you weigh it and compare it to your predictions. Most of the time you're off by some amount, and it often seems to be in that same direction: higher. So, as the weighed numbers come in, you are now plotting "actuals," compared to predictions, all against a hard limit. On Mars *Pathfinder,* our "predict" versus "actuals" showed a growth of over 60 percent from our first estimate. That kind of growth would generally be disastrous, but because we'd started with lots of margin, added to the margin with better navigation, and managed it carefully, we were actually able to add additional layers to the airbags and more rock protection for the lander late in the game.

—B.M.

that basically means the computer could do many things at the same time, instead of being limited to doing one thing after another.

Then Glenn adds the real reason why he believes *Pathfinder*'s software worked so well. "We had a really exceptional team of software developers here." And it was a remarkably small team. "Pam Yoshioka, David Cummings, Steve Stolper, Don Meyer, Carl Schneider, Kim Gostelow, Dave Smythe—those were the main developers. Greg Wells was our software quality assurance person. And we had many guidance and control analysts that worked on it, too, most notably Miguel San Martin and Sam Sirlin. It was a truly wonderful mix of people." But Flight System Manager Brian Muirhead thinks the real answer was that this team thought and acted like systems people first, software developers second. Glenn especially was always thinking about how the system was supposed to work, translated that understanding into operational code, then took it to the next step and put it on the hardware, making sure it did what it was supposed to, end-to-end. It was and still is a rare breed of engineer who can span the hardware/software/systems worlds, and they're worth their weight in platinum.

## CARL SAGAN

*A Pathfinder on Mars*

The exploration of space is one of the most exciting and important scientific adventures humanity can undertake. But by its very nature, it is also one of the most expensive. For now, planetary space exploration is something that can be afforded only by national governments (though many of us hope that will not be the case for long, witness the success of *Spaceship One*). And to be supported by those governments, space exploration must be supported by the people whose taxes fund it.

Carl Sagan was a master of garnering that broad-based public support for the United States space program and NASA in particular, and for the exploration of space in general. He didn't achieve this through debate or argument. Instead, he chose to share his immense passion and sense of wonder with the world, thereby helping a great many people previously unconnected to science to understand that scientific exploration is something that comes from the heart and not just the mind.

As a scientist, Carl Sagan worked with NASA since the 1950s, taking part in almost all the major missions to other worlds, from *Mariner 2,* to *Apollo, Viking, Voyager,* and *Galileo.* He was one of those who helped solve the mystery of the high temperatures of Venus, and then he showed the significance of that new knowledge by helping the public and politicians understand the related consequences of the greenhouse effect and nuclear winter on Earth.

As a popularizer of science, Sagan was seen by more than 500 million people in sixty countries around the world as the host of the award-winning educational television series *Cosmos.* The book adapted from that series became the most widely read science book ever published in the English language.

He was also a cofounder of The Planetary Society, the largest space advocacy organization in the world, and a tireless contributor to the ongoing quest to find indications of life elsewhere in the universe, from the radio-astronomy-based Search for Extraterrestrial Intelligence (SETI), to *Viking*'s chemical detection experiments to find signs of biological activity on Mars.

On December 20, 1996, as the *Pathfinder* spacecraft flew toward Mars, Carl Sagan died unexpectedly at the age of sixty-two.

NASA honored its deeply missed friend by officially declaring the *Pathfinder* lander on Mars the Carl Sagan Memorial Station.

When astronauts visit that station in person in the years ahead, they will leave their footprints on Mars, in part because Carl Sagan helped prepare the way for them to do so.

In 1995 Carl Sagan recorded a message for the future human inhabitants of Mars, which was included on disks carried by the two landers that were part of Russia's *Mars* '96 probe. Though the mission did not succeed, the message endures, and one day these words will be heard on Mars, as Carl intended.

I don't know why you're on Mars. Maybe you're there because we recognize we have to carefully move small asteroids around to avert the possibility of one impacting the Earth with catastrophic consequences, and while we're up in near-Earth space, it's only a hop, skip, and a jump to Mars. Or maybe we're on Mars because we recognize that if there are human communities on many different worlds, the chances of us being rendered extinct by some catastrophe on one world is much less. Or maybe we're on Mars because of the magnificent science that can be done there, that the gates of the wonder world are opening in our time. Or maybe we're on Mars because we have to be, because there is a deep nomadic impulse built into us by the evolutionary process. We come, after all, from hunter-gatherers, and for 99.9 percent of our tenure on Earth, we've been wanderers, and the next place to wander to is Mars.

But whatever the reason you're on Mars is, I'm glad you're there, and I wish I was with you.
COURTESY THE PLANETARY SOCIETY.

The Carl Sagan Memorial Station, Mars. COURTESY NASA/JPL.

This wonderful mix of people did more than deliver a robust software package for the Pathfinder mission. They actually set a new standard for software design for all future deep-space missions that followed.

Glenn laughs as he considers the irony of that accomplishment. "One of the things they told us at the beginning was 'Be bold.'" Indeed, part of Pathfinder's mandate was to break with tradition and find new ways of doing things. The software team did such a good job at that, their legacy has been the heart of the very successful Deep Space 1 and Mars Exploration Rover missions, and probably more to come.

## *PATHFINDER*'S LEGACY

A science-fiction fan for a navigator. An electronics test-bed engineer who "sucked" at electronics. Flight software developed by someone who had never done it before. The Pathfinder team seemed an unlikely group to make history. Its charter was to take risks but not fail. There were the rich experience and knowledge from past missions, but there was freedom in how to use that experience. There were no hard-and-fast rules other than the one that said "Do what has to be done for mission success—and don't be afraid to be bold."

The Pathfinder team followed that rule to the letter.

Though there were more than a few senior managers at JPL who felt that *Pathfinder* "would never make it to the launch pad," the mission fulfilled all its requirements with the unexpected extra benefit of electrifying the world and putting Mars on the front pages of newspapers, magazines, and television screens everywhere.

Not surprisingly, that level of success led to an inevitable question, much like the one Glenn Reeves faced in regard to his new approach to software design, and Pieter Kallemeyn about his team's precision navigation to Mars.

Did the Pathfinder team succeed because they were good or because they were lucky?

There's only one way to answer a question like that.

Do it again.

CHAPTER SIX

# "Standing on the Shoulders of Giants"

## Mars Exploration Roving

*Works of imagination have become so pervasive in American culture that the latitude of the government to satisfy them grows narrower by the day. Politicians are obliged by the nature of their jobs to satisfy public expectations, but the expectations that imagination creates grow more and more unattainable. Such a situation undermines government and generates public distrust with amazing speed.*

—HOWARD E. MCCURDY, *SPACE AND THE AMERICAN IMAGINATION*, 1997

## TWO PATHS, ONE DESTINATION

During the 1960s, to achieve President Kennedy's goal of landing humans on the Moon by the end of the decade, the number of people working for NASA and its contractors at the peak of the Apollo program reached 390,000, of which more than 35,000 were direct NASA employees.

But during the 1950s, many science-fiction movies had a different view of what it would take for humans to make that journey. More often than not, they featured spacecraft designed and built by a single scientist (male), usually in a remote desert base, and usually with the help of the scientist's pretty offspring (female) with whom the dashing pilot (male) would fall in love.

That 1950s dream of a privately built craft capable of taking humans into space was not achieved until June 21, 2004, when *SpaceShipOne* was launched from the White Knight carrier aircraft over California's Mojave Desert, and flown by test pilot Mike Melvill to an altitude of 328,491 feet, passing the accepted threshold of space by just under 400 feet.

In a 1950 movie, the privately built *Rocketship X-M* accidentally missed the Moon, but fortunately was able to make it all the way to Mars instead. COURTESY THE KOBAL COLLECTION.

The flight was a remarkable achievement on the part of aeronautics engineer and adventurer Burt Rutan, and Microsoft cofounder and philanthropist Paul G. Allen (who also founded the Science Fiction Museum and Hall of Fame in Seattle, of which ex-JPLer Donna Shirley is director). Though a small first step in comparison to the lunar voyages of Apollo, *SpaceShipOne* represents what some believe is, or should be, the future of near-Earth space travel: private industry.

One of the key reasons why Rutan's company—Scaled Composites, with only 135 employees—could finally accomplish private spaceflight in 2004 is because of the enormous government efforts of the past forty-plus years, and the wealth of information, technology, and experience those efforts have produced.

That's a pattern we've all seen before.

Christopher Columbus made history's first modern European voyage to North America, funded by the government of Spain. Today, anyone with a few hundred

The first real privately built crewed spacecraft—*SpaceShipOne*—just after pilot Mike Melvill made his historic flight into space. COURTESY MIKE OKUDA.

dollars can make the same journey in a matter of hours, using technology undreamed of in the fifteenth century—jet aircraft.

Inevitably, private spaceflight will advance in the same way, at first to take tourists on short, suborbital hops into space, and then into orbit. In generations to come, building on the cutting-edge NASA achievements of today, it's possible, if not easy, to imagine privately owned space stations serving as destinations for adventurous travelers. As of 2004, already two wealthy individuals—Dennis Tito (a former JPL engineer) and Mark Shuttleworth—each paid $20 million for a trip to the International Space Station aboard a Russian *Soyuz* spacecraft.

Looking even further ahead, given a century or more of continuing technological advancements in propulsion, engineering, and safety, there's no reason to doubt that someday even lunar excursions will be available to the well-funded tourist. And for those first commercial, nongovernmental flights to the Moon, we can be certain the company offering them will have far fewer than 390,000 employees.

NASA's first successful Mars landing mission—Project Viking—had taken eight years of planning and development. At its peak, it had a workforce in excess of

## RETURN OF THE MARTIAN JEDI

*Rob Manning Rides Again*

It was a dark time for the Rebel Alliance. On September 23, 1999, the Mars Climate Orbiter burned up in the atmosphere of Mars. On December 3, the wreckage of the Mars Polar Lander was likely strewn across its surface. Within a week of that failure, then Director of JPL Dr. Ed Stone declared it was "unlikely" the next Mars lander, Mars *Surveyor* (built on the heritage of the Polar Lander by the same company, Lockheed Martin Aerospace), would launch as scheduled in only sixteen months. By March 28, 2000, it was official—Mars *Surveyor* was "delayed."

But in the wake of the losses, Congress, in support, voted for additional funds for Mars exploration. Both NASA Administrator Dan Goldin and President Bill Clinton emphatically stated their support for NASA and JPL continuing the Mars program—except this time, the unofficial message was, whatever you do, don't fail again.

Even though JPL had funding, support, and an exceptional Mars launch opportunity coming up in 2003, its Mars program personnel were under pressure because they had nothing to launch in that window. In such a situation, who better to turn to than an engineer who had spent his childhood daydreaming through school, watching *Star Trek,* and being "mesmerized by the vision" of the epic science-fiction film *2001: A Space Odyssey?* Fortunately, that engineer had grown up to

become Rob Manning—one of the key people involved in JPL's last great Mars lander success, *Pathfinder.* Under the unique, fluid organizational structure of that mission, Rob had come aboard *Pathfinder* as an electronics engineer fresh from the *Cassini* Saturn probe and ended up as the engineer responsible for, among other things, the mission's complex interplay of systems for Entry, Descent, and Landing on Mars.

Looking back, Rob really doesn't remember when he set his course on a career in engineering. "I think you make up your mind early when you're a kid. It's not a predisposition, or a predisposed talent—it's just an interest that turns you on. I never met an engineer until I got to college."

Rob grew up a "navy brat" on Puget Sound, but his first outings into space came in the second grade by way of the *Space Cat* children's book series. "This cat somehow goes along for a ride with his astronaut friend, and they visit *all* the planets. That was a big thing."

As it turned out, Rob's active imagination became an important part of his early experience with engineering—long before he knew what the term meant. In seventh grade, encouraged by his father, Rob explored his interest in airplanes. "Although I didn't have access to flying," he recalls, "I had access to touching planes and looking at planes, and I also had access to the *books.* I pored over books about airplanes and learned piloting and navigation skills, and when I was fourteen, I passed the written flight exam. But I didn't have the money to fly."

The same situation arose when Rob ex-

plored his interest in radio. "I used to order ham radio books about how to build your own. But the books had to substitute for the real McCoy because I didn't have money for the parts."

But for all his outside interests, Rob found school very boring. "Quite frankly, what I was doing in school was staring out the windows all day long, daydreaming."

Still, there was a strong, practical side to Rob. Growing up in farm country, he worked every summer starting at age twelve, operating combines and driving tractors and trucks—legal on a farm if not on public roads. After moving from Puget Sound, he attended an agricultural high school. "There was virtually no science or math to speak of," he remembers. Then with his wide, trademark smile, he adds, "But on the other hand, it had a plastics shop, a small gas-engine shop, a machine shop, and a welding shop. Welding is a lot of fun for a kid!"

After high school, Rob attended Whitman College in Walla Walla, Washington. Though it was a liberal arts college, it offered a program in which students could attend Whitman for three years, then take an additional two years at Caltech. Rob majored in math and physics, transferred to Caltech, and never looked back. Even though "engineering scared the bejeebers out of me," he decided to study electrical engineering "because it was the most mysterious." The year was 1979, "when computers and electronics were really exploding, and I figured if I didn't learn this now, I'll never learn it. So I zoomed in on that and found it to be really interesting."

two thousand, including personnel from NASA, JPL, and private industry, chiefly the Martin Marietta Corporation.

Twenty-one years after the expensive but very successful Project Viking, the Mars Pathfinder mission took less than four years of planning and development, and using a peak workforce of about 275 at JPL. Admittedly, Viking was a larger, more complex mission that called for two landers, two orbiters, and a full suite of scientific instrumentation, while *Pathfinder* consisted of a single lander, a micro-

Rob first applied to JPL as a student, and became an APT—academic part-time employee. His first job was to oversee the manufacture of the computer memory chips to be used in the *Galileo* probe to Jupiter, and there was no doubt he was hooked on his new career. "Why am I here?" he asks. "I'm here because I think spacecraft and planets and space exploration are really cool!"

Eventually, Rob came to work on the *Cassini* probe, specializing in fault-tolerant software—that is, software that has the capability to detect errors during flight and either correct them, work around them, or put the spacecraft into a "safe state" and then phone home for instructions. While Rob was on Cassini, the Mars Pathfinder program began, and Brian Muirhead found himself in need of an electronics and software engineer—and not just any engineer. Brian went to the section manager of JPL's computers and software and asked who was the best they had. The answer was Rob Manning. The section head also added, "And you'll never get him—he's on Cassini." But over a lunch conversation, during which Brian told Rob the Pathfinder plan and Rob called the plan "nuts," Rob became hooked. Within a few weeks, he transferred from Cassini to Pathfinder, to be the flight system chief engineer.

But, several months into the project, with the computer and software engineering no longer at the top of the list of concerns, Rob found himself drawn to the most critical area and another compelling interest: the design, development, testing, and operation of the craft's Entry, Descent, and Landing system.

After *Pathfinder,* Rob's next Mars assignment was mission manager for the Mars Sample Return, scheduled to launch its first sample-gathering rover in 2003. Unfortunately, this ambitious mission was destined to be canceled, another casualty of the '98 losses.

But after those losses, Rob's wealth of experience put him in a perfect position to help suggest the next step JPL could take within NASA's newly reorganized Mars Exploration program.

Rob clearly recalls going with Mark Adler, then chief engineer for the Mars Sample Return mission, to a meeting in April 2000 with Firouz Naderi, the new manager of JPL's Mars program. At the time, NASA and JPL remained committed to launching two Mars missions every twenty-six months—ideally, an orbiter and a lander at every launch opportunity. But no one at NASA or JPL was sure whether, given the late date, anything could be done fast enough to meet a 2003 launch.

Some engineers had suggested JPL not *develop* a new lander—merely use the plans for one that had already worked and "rebuild" a new version of Mars *Pathfinder.* But Rob and Mark were suggesting a variation on the theme—using the *Pathfinder* airbag landing system to deliver a bigger, more capable rover to Mars. Of course, that would require designing, developing, and testing a totally new rover in less than three years instead of the usual four to five years.

Rob laughs as he recalls the results of that first meeting: "We got kicked out."

Still, the idea had enough merit that Rob was assigned to lead a systems-engineering study of their proposal.

By May 2000, the Mars Mobile Lander mission was one of two proposals put on the fast track for consideration by NASA. The lander concept review was now being led by JPL's Peter Theisinger. The second proposal, for an orbiter mission, was being developed with Lockheed Martin Astronautics. In early July, both proposals were delivered to NASA Headquarters. It was up to NASA Associate Administrator for Space Science Dr. Ed Weiler to choose between them.

In the end, the decision was tough. Though both lander and orbiter missions were estimated to have the same cost and deliver excellent science, plans called for there to be at least four other orbiters in operation by 2003. (The projection was close: three orbiters in operation in late 2003 and still in operation today are NASA's Mars Global Surveyor and Mars *Odyssey,* and the European Space Agency's *Mars Express.* A fourth orbiter, the Japanese spacecraft *Nozomi,* malfunctioned on its way to Mars.) Since NASA hadn't had a Mars lander since *Pathfinder,* and because the launch opportunity was so favorable, Dr. Weiler chose the lander.

Rob Manning, that daydreaming kid, was heading back to Mars.

And this time, because of a simple question asked by NASA Administrator Dan Goldin, Rob and the rest of the Mars Exploration Rover mission team were going to have *twice* as much fun.

rover, and a limited science payload. But even accounting for differences of scope, *Pathfinder* was still a much leaner operation—exactly what one would expect from an undertaking able to draw upon the knowledge gained from the first Mars landings, and dedicated to the principles of FBC—faster, better, cheaper.

Designed to launch seven years after *Pathfinder*, the twin Mars Exploration Rover missions of 2004 had a larger combined workforce than Pathfinder, even exceeding the Cassini mission peak of six hundred at JPL alone.

Why?

Detailed discussions about the new approach to Mars exploration involve as many opinions as participants. But all concerned agree that the underlying reason for the change to an increase of cost and personnel was the painful loss of the Mars Climate Orbiter and the Mars Polar Lander in 1999. (They're referred to as the '98 failures because NASA often identifies missions by the year they're launched—though the Mars Polar Lander was launched January 3, 1999).

NASA's intentions had been to carry out both '98 missions according to the faster, better, cheaper model that had worked so well for *Pathfinder*, even though the commitment to those missions was made before Pathfinder had even gotten off the ground. All involved understood (or should have) that following that model entails accepting a significant level of risk in exchange for the benefits of innovation and low cost.

Had only one spacecraft failed, it might have been viewed as the price of the "faster, better, cheaper" approach, though hard questions would have been asked of the one craft's failure.

But *both* spacecraft were lost, which meant no science was returned—this price seemed too high.

The review boards and investigations that followed were thorough, and NASA and, especially, JPL made many changes in how they did business. Some people would argue that the 1998 missions weren't really run according to faster, better, cheaper principles. Others would say that faster, better, cheaper isn't appropriate for certain types of missions. With so many sincere and dedicated people involved in the debate, it's not within the scope of these pages to draw ultimate conclusions.

But there's no denying that in the aftermath of the '98 failures, when it came to the exploration of Mars, the idea of "acceptable levels of risk" was something NASA was no longer willing to consider. And such risk avoidance would mean the next missions would no doubt require more money and more people to achieve success.

Before the '98 failures, the next two missions to Mars were already being developed and built for the 2001 launch opportunity. The lander mission, Mars *Surveyor*, was based on the same design as the lost Mars Polar Lander. Now it was canceled, and the fully built spacecraft put into storage with an unknown future. (See chapter 4 for details of the upgraded Mars *Surveyor* lander flying as the Mars *Phoenix* lander in 2007.) The orbiter mission, Mars *Odyssey*, proceeded, but under heightened scrutiny. (On May 22, 2004, *Odyssey* completed its *ten thousandth* orbit of Mars, continuing to provide unique new information about the chemical and mineralogical make-up of Mars, as well as mapping the presence of unexpectedly large amounts of hydrogen, which some researchers think could indicate the presence of underground ice.)

But back in early 2000, the Mars program was still scrambling—past the 2001 *Odyssey* launch, nothing was certain.

Except physics.

Because of the close approach of Mars and Earth—the best opposition in more than 59,000 years!—the 2003 launch opportunity represented the most favorable NASA had had since the space age had begun.

How could NASA *not* launch a lander mission at that time? Yet with the cancellation of Mars *Surveyor*, there was nothing on the drawing boards.

Or was there?

After all, *Pathfinder* had been a complete success, and JPL still had all the blueprints.

*Could Pathfinder* fly again?

The decision was made less than five months after the Mars Polar Lander was lost. In April 2000 JPL proposed a new Mars mission for 2003—one that was as close to a guaranteed success as possible, because the Lab intended to go the tried-and-true route and build a duplicate *Pathfinder* lander to carry a new rover, derived from *Sojourner*, but with a much larger science package.

The idea made so much sense that when NASA chief administrator Dan Goldin heard the proposal, his first question was, "Why aren't you proposing two rovers?"

Why not, indeed.

Virtually overnight, the Mars Exploration Rover project began: two landers, two rovers to be launched in June 2003.

To achieve that almost impossible goal of getting to the launchpad in slightly more than two years, it was clear that the team responsible was going to have to be the best that JPL could field.

And to have the best team, they were going to need great leadership.

Fortunately, they had it.


## THE ONLY PEOPLE

On January 3, 2004, this is how Peter Theisinger started his day.

"I got up that first Saturday when *Spirit* was to land, and as I was shaving I looked was in the mirror and I said, 'Okay, when I look in this mirror tomorrow, the world will be different.' I did not know how it 'will be different,' but it *will* be different . . . because we will hit Mars tonight. One way or the other, we will hit the planet's surface."

Pete was the project manager for the Mars Exploration Rover mission. That night, at the press conference following the successful landing of *Spirit*, millions of people around the world saw a happy, smiling Peter Theisinger, his cap of short white hair as distinctive as Rob Manning's beard or Matt Golombek's laugh.

But that morning Pete wasn't yet JPL's latest star—he was the outwardly calm engineer who had been given twenty-seven months to complete what should have been at least a four-and-a-half-year assignment. And even without the benefit of hindsight regarding the fantastic success of both Mars Exploration Rovers, Pete Theisinger was the right choice to lead one of the toughest missions JPL had ever undertaken.

Taking well-deserved bows after *Spirit*'s successful landing on Mars, on behalf of all the workers who made it possible. From left to right, NASA Administrator Sean O'Keefe, NASA Associate Administrator for Space Science Ed Weiler, Director of JPL Charles Elachi, MER Project Manager Pete Theisinger, Deputy Manager Richard Cook, and Lead Engineer for Entry, Descent, and Landing Rob Manning.

Pete's a quiet, unassuming fellow who at first doesn't think there's anything particularly telling about his early years that might have indicated he was heading for a career in space exploration. Slowly but surely, though, as his story unfolds, all the pieces of the puzzle emerge. Like so many others of his colleagues at JPL, Pete was destined for space from an early age—he just didn't know it at the time.

Pete looks upon his early interest in amateur radio and assembling Heathkit electronic hobby kits as "the first manifestation" of his involvement in engineering. Other than that, though, he says, "I'm very atypical, I bet. I was never really a science-fair junkie. I didn't do that kind of stuff." Like most boys his age, Pete made models of planes and tanks, but when it came to his first real car, he notes, "I was never mechanical, never tried to fix my car." He adds with a laugh, "I'm a klutz, you know."

But space was part of his environment, so much so that he didn't even seem to notice it. Not only was he an avid reader of science fiction, his father was an electrical engineer in the aerospace industry. One of his jobs was working for the space division of North American Rockwell during the Apollo project, as part of the ground-support team for the *Apollo* command module.

During the days of *Apollo*, Pete was a student at Caltech. In his usual, unassuming way, he attaches no great significance to the fact that he became a student at one of the most prestigious universities for science and technology. "I lived close, for one thing. I lived in Southern California at the time. But I was also very close to the top of my class in high school, and in those days, if you were at top end of your high school class, you tended to see math and physics and chemistry as exciting topics, and history and English and those things as being perhaps a little less exciting." With another laugh, he adds, "And things like economics and anthropology and those kind of things—not even on the event horizon."

January 3, 2004. The first 360° panoramic image returned from the rover *Spirit*, showing its landing petals, deployment ramps, deflated airbags, and the Martian horizon. COURTESY NASA/JPL.

Pete's first direct encounter with the space business took place between his sophomore and junior years of college, when he spent the summer with his parents in Florida.

"When I got to Florida I didn't know anything. I didn't know anybody. And I was looking around for a job. It turned out there was a gas station just opening about a mile and a half from where I was, so I went to work there. Just pumping gas." Pete points out that his job was strictly limited. His lack of mechanical aptitude kept him from even adding oil to cars. "But pump gas, I can handle."

One of the services the gas station offered was a car wash provided by a youngster in the neighborhood. "We would charge a dollar for the car wash and we would give him fifty cents and we would keep fifty cents. We supplied the sponges and the soap and the buckets of water, and he would supply the labor.

"So one day this lady drives in with her large car. And she's sitting in the office with me while her car's being washed, and she says, 'You don't sound like you come from here.' I say, 'I come from Southern California.' She asks, 'Where do you go to school?' I say, 'I go to Caltech.' And she says, 'What are you doing at a gas station?'"

Pete laughs as he remembers his reaction to the bluntness of her question. "I said, 'It's the only job I could get.' She said, 'Well, maybe my husband can help you out.' Then she drove off.

"The next day, the same car drives up with this guy. He leans over and says, 'Are you the guy from Caltech?'" More laughter. "Turns out, the woman's husband works for the Bechtel Corporation." Bechtel is a prominent construction company. "They had the contract at the Cape for doing the gantry and pad modifications between launches. So every time they had a launch, Bechtel had to repair the damage and change the configuration for the next type of booster."

The fellow told Pete that their workload had exploded, and "He was looking for a go-fer. He asked, 'Do you want a job?' I said, 'Sure!' He said, 'Can you start on Monday?' Well, this was on a Saturday, and I said, 'How will I let these people know?' He said, 'What have they ever done for you?'"

So Monday morning, Pete started working at the Kennedy Space Center as the Gemini program was under way.

Pete returned to Caltech in the fall, but often spent his vacations with his parents in Florida. He remembers the highlight of one vacation trip being an unexpected call from one of his father's friends who worked in what the military calls "black programs." That is, secret projects.

One day, the friend called Pete and said, "Let's go play golf tomorrow." Pete remembers how puzzled he was by the call. "He had never played golf with me in my entire life. So, I said, 'What?' And he said, 'Trust me. Let's go for golf tomorrow. I'll pick you up at six A.M.' I said, 'Ohhhkay . . . '" But Pete had no idea why the invitation had been extended.

"We went up to a little pitch-and-putt range and we just kind of played along, until he looked at his watch and he said, 'Let's take a walk, okay?' And across the street you could see the big Titan pads." The Titan launch pads were at Cape Canaveral Air Force Station, which is literally across the river from the Kennedy Space Center. Pete smiles as he remembers the wonderful experience his father's friend made possible—witnessing the launch of *Gemini VIII*, carrying astronauts Neil Armstrong and Dave Scott.

By that time, there was no question about it, Pete was hooked—but he still hadn't quite realized it.

The *Gemini VIII* launch was on March 16, 1966. Just over a year later, Pete graduated from Caltech with a degree in physics and was accepted by the University of Michigan for graduate studies. For his 1967 summer job, he applied to JPL and was able to put his extensive knowledge of physics to work by . . . watching Teletypes on the graveyard shift.

The *Mariner 5* probe to Venus had just launched on June 10, and in those days before computers provided round-the-clock monitoring, Pete's job was to watch the Teletypes print out information returned from the probe. If anything seemed to be going wrong, he was to open the phone list and call the right person. "That was the extent of my control," Pete says. "I did it for about three weeks. By then, the mission was going very well so they had me do some other things."

It was in the course of doing those "other things" that Pete finally *realized* he had been hooked.

"About halfway through the summer, I said, Well, this is fun and grad school is many years to get a PhD . . . I wonder if they'd let me stay?

"I went to the right people and they put me to work on the original *Voyager*."

Though today we know the *Voyager 1* and *2* space probes as those that accomplished the amazing Grand Tour of the outer planets and are now at the edge of our solar system, never to return, the Voyager program Pete began working on had begun in 1960, as an ambitious plan to send combined orbiter and lander missions to Venus and Mars.

But by the summer of 1967, NASA had been spending money on *Voyager* studies for seven years. Between the escalating costs of Project Apollo and a new call for proposals for human flights to Mars from the Manned Space Center in Houston (now the Johnson Space Center), Congress had had enough.

Thus, about a week after Pete Theisinger began work on what promised to be his first trip to Mars, "Congress canceled the original *Voyager*, and I went to talk to my new bride and say, 'You know, we may have some difficulties . . .'"

But even more laughter accompanies the memory of that sudden reversal of fortune, because it had a happy ending. "When I went to work the next day, my boss very quickly got me in his office and said, 'You are not to worry—we will find you something to do. You are not at risk..'"

Over the next several years, Pete worked on a variety of advanced studies, including what would become the *Voyager* Grand Tour. Up until the success of *Spirit* and *Opportunity*, the Voyager mission gave Pete one of his most memorable experiences at JPL.

"Just before *Voyager* arrived at Jupiter, I got a phone call from someone who said, 'Go to the press conference today.' I went down to the press conference and for ninety days, we had been seeing pictures every forty-eight seconds of Jupiter with the little moons going around, and now we were getting really close, only two or three days out. So we had got in just close enough that we were able to take the first full-color mosaic picture of Io." (Io is one of the four largest moons of Jupiter, about the size of Earth's Moon, and the most volcanically active body in the solar system.)

"It was the famous picture of Io, that some people describe as a rotten orange attacked by an ice pick. Someone simply got up and said, 'We want to show you what we got yesterday.' And they opened up the theater screen in the auditorium and just showed this picture—didn't say anything, just showed the picture.

"The entire audience just gasped— a very audible gasp. And I said to myself at the time, I *know* why I do this. The 'psychic' return is phenomenally high, and we *are* addicted to it."

After *Voyager*, Pete then worked on the *Galileo* probe to Jupiter, and after that, left JPL to work for one of *Galileo*'s outside contractors.

Pete's sojourn in the outside world lasted three years. Like Glenn Reeves, he found there were financial advantages to stepping outside JPL's system of promotions and pay raises, but when the company he worked for became primarily a defense contractor, Pete realized where his passion and his interest in space truly lay.

"It wasn't the fact that it was military versus civilian," he explains. "It was the fact that the business was different. The actual work would be different. The assumptions, the acquisition strategy, the risk paradigm, all of that stuff would be different.

"Also, my experience at this point, for a decade and a half, was in an interplanetary world." Since defense contractors deal much closer to home, "I would be going into an Earth-orbit environment, spy satellites, intel—a different world, different technology, different expertise. I'd be very much a junior person in that environment.

"A rotten orange attacked by an ice pick." The Jovian moon Io as seen by the *Voyager 1* probe, which so impressed Pete Theisinger. COURTESY NASA/JPL.

"So I thought, I don't want to do that, and I decided I should go back to the Lab. And I did."

The year of Pete's return was 1983, and over the next decade he steadily rose through the ranks, taking on new jobs and new responsibilities. A decade later, he took on a major assignment as project engineer for one of JPL's greatest success stories—the Mars Global Surveyor (MGS). In a touch of foreshadowing, MGS was a project that came about in a similar fashion to the Mars Exploration Rover (MER) mission. It was a fast-track, low-cost mission intended to quickly recover the mission objectives of the Mars Observer, which was lost during its final approach to Mars on August 21, 1993.

It was through MGS that Pete learned an important lesson, one that he kept in his mind, and the mind of his team, throughout the MER mission.

"I'd had a conversation with Mitchell Troy at JPL, about the Mars Global Surveyor program, which happened reasonably quickly after the loss of Mars Observer. And he said something that I found always intriguing, which was: 'They will forgive you for being unlucky, but they will not forgive you for being stupid.'"

Pete says that became the guiding principle for the Mars Exploration Rover project: "Don't be stupid."

As mentioned, NASA's missions are developed according to the way four key priorities are ranked—what planners call a "paradigm." Those four priorities are Cost, Schedule, Reliability (including safety), and Performance. At the time of the Apollo race to the Moon, the order of importance for those priorities usually came out with Performance and Reliability in the #1 and #2 slots. Human lives were at stake, so spacecraft and their related equipment and procedures had to work flawlessly, as well as have exceptional backup capabilities in the event of the unexpected. Schedule came next—not only to achieve President Kennedy's timetable, but also to beat the Russians. And finally, Cost came in at #4, the lowest priority— reflected in the constantly increasing budget.

Mars Pathfinder turned this paradigm around. Cost became number one. If it even *looked* as if the project would go over budget, it would be canceled outright (at least that's what the project's team believed). Schedule was a close second. Launch windows for the Moon come around twice a day. Miss one, and there will be another available almost as soon as the computers can be reprogrammed. But usable launch windows to Mars come around once every twenty-six months or so, and last only about twenty days. Miss a Mars window, and three years of work and hundreds of millions of dollars can be lost.

Reliability was in the #3 slot, with Performance bringing up the rear. All the Pathfinder mission had to accomplish was a safe landing that would allow its rover to drive off and its camera to take a high-resolution panoramic image. The mission success criteria required seven days of rover operation and thirty days for the lander, but if the lander and rover were to operate for only one day, the mission would probably still be a success. (In fact, they lasted 87 days.) Indeed, Dan Goldin, the chief proponent of "faster, better, cheaper," proclaimed *Pathfinder* a success a day *before* it had landed, if only because the team had demonstrated the ability to get a spacecraft designed, built, and launched under the schedule and budget restraints of a Discovery mission.

But with the faster-better-cheaper paradigm on the ropes after the '98 losses, and NASA needing a success at Mars, the Mars Exploration Rover mission required another paradigm shift.

Pete Theisinger has a succinct way of describing why. "It became a success-driven paradigm. After the loss of the '98 missions and then the restructuring of the program that occurred in 2000, MER was perceived to be a risky mission [as if landing on Mars could ever be routine]. But a risky mission for which the payoff was very good."

Now a new priority found itself in the top position. "The thing on MER was, the schedule dominated *everything*. You'd make decisions, and then you might realize, Well, I should have studied that in more detail. But the fact is, you *had* to make a decision, and if it turned out to be not-so-good a decision, then we'll find that out and we'll fix it later."

In fact, Schedule was so important, so critical, that Pete doesn't even call it number one. "Everyone understood the stakes were high, from both the planetary program standpoint and then later, as it transpired, from a NASA standpoint. But the mantra was always: We're going to do this the way we *know* how to do this work. Success [Reliability] was number one. Performance was number two, but Schedule was number zero, because we *had* to get off on a very short schedule. We would try to keep Cost in the bag, but, if Cost had to go, then Cost had to go. And that's the way it played out." For the record, that decision to pay whatever it cost for success wasn't Pete's. It had to be approved by NASA Headquarters.

There were times when Pete was offered other opportunities to cope with the demands of his schedule. "I was invited, several times, to give something up from the mission-content standpoint if it would make my life easier, or if it would give me a better chance for meeting the schedule. The question always was: 'Can we help you with the schedule if we give up X, Y, or Z?' But we never were in a situation where we felt there was a need for anything like that."

The legacy of previous missions was one of the main reasons why it was possible for Pete to maintain the full scope of MER's mission objectives, despite the severe constraints of schedule.

"Some of the initial development stages did not have to be repeated. We were able to use some inherited components, rather than new ones requiring heavy development." Typically, one of the highest risk areas is the development of science instruments. But because the MER instruments had already been developed for the canceled '01 Mars *Surveyor* lander, "We had a mature payload."

In addition to the payload, the MER system design, configuration, and mission plan—up to landing—could also be largely inherited from previous missions. "*Pathfinder* had done a very successful Entry, Descent, and Landing with its new architecture. Not only was the architecture and basic design done, but also all the thinking that went into the validation program. We knew the tests we'd have to do; the simulations we'd have to do; the way to do the simulations; all those things."

But payload and architecture still aren't the most important ingredients for mission success—people are. And once again, past missions, especially Pathfinder, provided much of the legacy.

"We used a bunch of people who had been through this process before, so the second time through, they understood where they didn't do it quite the way they would like to do it the first time.

"As I've said to my MER team: 'You are not the *best* people in the world to be doing this; you are the *only* people who *can* do this.'"

He was right.

Some of the team members who contributed to the spectacular success of the Mars Exploration Rover missions. COURTESY NASA/JPL.

## ELVIS LIVES

Way back in chapter one, in describing a tense moment during *Pathfinder*'s final approach to Mars, we made mention of computer "fill packets"—empty spaces in strings of zeros and ones that computer programmers would fill up with self-described "geek humor." One part of every fill packet on *Pathfinder* was "Elvis Lives." And in one sense, it's partially right. Though Elvis is no longer present on Earth, his sideburns live on, carefully tended by Adam Steltzner, one of the more colorful members of the MER team, with a personal trajectory to Mars that's truly unique.

To begin with, when asked when he first thought that he'd like to work in space exploration, Adam replies, "Two weeks before I started work at JPL."

Not that Adam didn't always have an affinity for astronomical topics. "When I was in nursery school," he recalls, "and they went around the room and asked, 'What do you want to be?' I answered that I would like to be the sun and shine down upon everyone."

Adam laughs at that answer, wondering if he might be revealing too much about his past. But as he puts it, his childhood was different from those of most others at JPL, because "I was raised by beatniks. So the arts were focused on and I never did well in school at all. I failed multiple grades. I only passed high school on my charm—which was legendary," he adds, quite unnecessarily.

But the other reason he didn't fail high school is a good indication of his future academic success: "the fact that I buckled down enough—because I did not want to be a high school dropout."

For Adam Steltzner, it all comes down to motivation.

After high school, he focused on his main interest—music, playing bass guitar and drums. "Mostly rock and roll," he says, "or new wave, as it was called in those days." He even studied jazz at the Berklee College of Music in Boston, whose alumni include Quincy Jones, Melissa Etheridge, film composer Alan Silvestri, and *The Tonight Show*'s bandleader, Kevin Eubanks.

However, Adam stayed for less than a year. Many explanations are possible, ranging from his self-described inability to plan, his fear of failure, and the fact that his family was sufficiently well off that money was not an issue.

But in Adam's case, money did not buy happiness. "So somewhere around 1984, I decided that I wanted something different. I decided I wanted to apply myself to something. I'd never really found that something that I wanted, so that was going to be my thing. I was going to try something and maybe fail. And what I decided I'd try and maybe fail at was school."

At the time Adam was going through this reevaluation of his life's direction, he was playing music at a club in San Francisco. And whether through boredom or by accident, "I would be playing the shows and I would notice that a different set of stars were in the sky on my way back from the gig than were in the sky when I was on my way out to the gig."

So Adam dropped in at the City College of San Francisco and decided to sign up for a course in astronomy, to learn about those changing stars.

However, to take the astronomy course that was offered, he was required to take a prerequisite course in introductory physics.

He took that course, and in one moment, one simple equation written on the board, changed his life.

The course was taught by an instructor who had earned his doctorate at Caltech. To prepare his class, he began with a review of basic algebra and how it's applied in physics. "I can remember exactly this is what he said," Adam recalls. "F is MA, and then he says, A is F divided by M."

---

The instructor was showing the relationship between Force, Mass, and Acceleration (also known as Newton's Second Law of Motion). In the first equation, to know the Force acting on an object, merely multiply the Mass of the object times the Acceleration it's undergoing. And since all the elements of an equation can be reshuffled to reveal new relationships, Acceleration can be calculated by dividing Force by Mass.

---

The bottom line is that if you know any two quantities in this simple equation, mathematics can predict the third one.

Adam's reaction to that revelation was "Whoa!"

For whatever reason, at that moment, in that classroom, looking at that simple, very basic mathematical relationship written on the board, Adam Steltzner experienced what generations of those drawn to science have always shared: the power of prediction, the astounding realization that the physical basis of nature is not arbitrary, but can be defined, measured, predicted, and *understood*.

As Adam puts it, "I had found religion."

And with that discovery had come what he had been missing for so many years—motivation.

"That was 1985. I quit the band and I dug in. And I *seriously* pounded the competition." Unlike the other students in his classes, who were attending school because it was what was expected of them, Adam *knew* why he was there. But then he had another realization. Physics was aesthetically pleasing, so to him, it felt like a kind of art. And art wasn't what he was looking for.

"I didn't want to do just another art. I'd done music." But fortunately, there was another field closely related to science. "This engineering thing," Adam remembers thinking, "It's like a practical application of physics. You might be able to make a living at it, which would break this long-standing tradition of not working that has been in my family for years!

"I liked that idea. A *job*. Something you could actually work at. Someone might pay you for it."

Motivated, driven, Adam worked hard at City College, moved on to the University of California at Davis, and then to one of the ultimate destinations for the study of engineering, Caltech. There, he earned his master's degree in a year, immediately applied for a job at JPL—and met with disappointment.

The first manager to interview Adam was known to be "famously cantankerous," while Adam was supremely confident, talented, and competitive. Given that volatile mix, the interview became a debate, and Adam argued positions in opposition to the interviewer. At the end of thirty minutes, Adam recalls he was told, "You're not ready for JPL and thanks very much. You can find your own way out."

Adam can laugh about it today, but at the time, "I was crushed. It was a devastating, very difficult experience." Given the preceding few years of Adam's academic success, he also found it an experience that was novel, almost surreal. Failure was not something he was familiar with.

However, he faced that initial defeat with the proper strategy.

"I picked myself back up. That's the only essential lesson to learn—to go forward, endlessly, and pick yourself back up. Because if you can't pull yourself up from failure, you're not trying."

Adam staged his second assault on JPL. "I looked at the JPL phone book, circled each manager in divisions that seemed like they'd be a good fit, sent résumés, and followed with a cold call."

The next manager Adam spoke with was in the Applied Mechanics section. The manager remembered getting Adam's résumé and said he had passed it on to another manager. Adam dutifully phoned that manager, but once again faced failure when told, "Adam, I have a lot of résumés on my desk here, I have a list of the ones that I'm interested in, and I don't find you on this list. But good luck! Keep in touch."

Adam was stunned. "Because I look pretty damn good on paper. There *aren't* that many people out there! So I'm thinking, *Who* is getting *hired* at JPL?! I mean, *who are these people?!*"

But then, "Ten minutes later, the manager phoned me back." He had done some digging and realized he hadn't seen Adam's résumé after all and asked him to come in for an interview the very next day.

Success at last.

Adam then followed a somewhat more traditional path at JPL, working on a variety of projects, including *Cassini* and a short-lived but fascinating mission called Champollion, led by Brian Muirhead. The ambitious *Champollion* mission was intended to rendezvous with the comet Tempel 1, release a lander that would attach itself to the comet, take core samples and analyze them, looking for the building blocks of life that are thought to be locked up in these 4.5 billion-year-old time capsules. The technical and funding complexity of the mission proved too daunting for the time, though Adam's technical skills and tenacity in developing and testing the "harpoons" needed to anchor a lander to the completely unknown surface of a comet drew him to the attention of some key people who would help guide him in the future.

(Another comet mission, *Deep Impact*, will launch in December 2004 and rendezvous with the same comet in July 2005. The Deep Impact probe will then release an 820-pound projectile that will collide with the comet at 23,000 mph. This unique experiment will create a crater, possibly as large as a football field, that will enable scientists to understand something about the structure and composition of a comet.)

Eventually, Adam came to work on the Mars Exploration Rover mission. The project came together blindingly fast and required so many people to meet its accelerated schedule that, as Adam explains, "The managers would just walk down the list of personnel. If someone wasn't critical to another project, they were likely to be headed toward MER."

At the time, Adam *was* involved on another project, studying a technique called aerocapture for a future Mars orbiter. (The basic idea is that to slow down a probe enough to enter orbit of Mars, there are two techniques. One is to have the probe carry enough fuel to use rocket propulsion for braking. The other, less massive but riskier option is to have the probe fly deep enough into and out of the Martian atmosphere to slow down and be captured by Mars's gravity.)

But the '05 mission was a long way away compared to MER. And three to four months into the mission planning, as Pete Theisinger and his team realized that MER was going to be more complex than merely building two *Pathfinder* duplicates, the word went out that MER needed additional expertise in Entry, Descent, and Landing. So Adam was reassigned from studying aerocapture to becoming the mechanical lead engineer for Mars '03 EDL.

The assignment thrust Adam into a pressure cooker, and he thrived on it. EDL components were manufactured by contractors across the country, so he traveled over 100,000 miles a year. At the same time, his wife worked in New York, meaning their schedules matched only once or twice a month. But as Adam says, with the support of his wife, "The job demanded just what I do well."

By the time Adam joined MER, the idea of a near-identical rebuild of *Pathfinder* had proven infeasible. The *Pathfinder* architecture would be followed, but all

the basic pieces would need to be changed—some a little, some a lot. Each of the two MER spacecraft would travel directly to Mars and descend without orbiting. A parachute would be deployed at supersonic speeds. The heat shield would be jettisoned. The lander would be a tetrahedron that dropped from the backshell on a bridle. Seconds before impact, airbags would inflate, rockets would fire to reduce the descent rate to zero, then the bridle would be cut, the backshell soar away, and the airbag-enclosed lander would bounce across the Martian surface until it rolled to a stop, the airbags deflated, and the lander petals opened to free the rover inside. It's just that easy!

Except, because the rovers for this mission would be bigger and heavier, the lander would need to be stronger, which meant the overall spacecraft at entry to Mars's atmosphere would be heavier, and that meant *everything* had to be

On July 4, 2005, eight years after Pathfinder, another Discovery mission, the *Deep Impact* probe, will launch an 820-pound "impactor" at comet Tempel 1, to collide with it at almost 23,000 mph. The mission's goal is to reveal the comet's interior by blasting a crater the size of a football field. COURTESY NASA/JPL.

## MOVE AND WAIT

*Rover History, Part One*

JPL has been researching the use of rovers to explore other worlds since the early 1960s. The first major development occurred in 1963, when the Lab contracted with General Motors to produce the *Surveyor* Lunar Rover Vehicle (SLRV). This vehicle was a prototype for a rover intended to land on the Moon as part of the Surveyor robotic spacecraft program. Its purpose was to survey landing sites for subsequent Apollo missions. However, the SLRV was never flown because the *Surveyor* landers showed the Moon's surface to be firm enough for astronauts and their equipment to operate.

Plans called for the lunar rover to be operated by a technique simply known as "move and wait." First, the rover transmits an image from its forward camera to ground control. On Earth, the human operator selects a steering angle for the rover's wheels, then triggers a command for the rover to move forward one wheel-circumference (about five feet). Then the rover transmits an image from its new location and the cycle repeats. This technique

The first autonomous rover to operate on another body—Russia's *Lunakhod*. COURTESY NASA

redesigned—bigger parachute, stronger airbags, bigger rockets, even a slightly different shape of aeroshell to hold the larger vehicle.

And it all had to be accomplished in less time than *Pathfinder*.

The result was inevitable, Adam says. "During development, we had big problems on the airbags. We had big problems on the parachute, *and* the descent-rate limiter. Actually, there were really very few subsystems that I was 'mother-goosing' that didn't have some serious problems. Though," he adds, "Lockheed Martin's work on the aeroshell went fairly smoothly."

Given the unforgiving schedule, Adam adopted a strategy for dealing with the problems he faced. It wasn't the most cost-effective way to proceed, but as Pete Theisinger had directed, schedule was paramount and if cost had to go, cost had to go.

"What we would tend to do when things were questionable, I'd send two or three paths forward." As an example, he describes one of the most potentially disastrous problems the mission faced—even more so because it arose very late in the dewas possible because of the short round-trip light time from the Moon—less than three seconds. Extensive testing of this approach to controlling a rover was conducted at JPL in the 1960s and early 1970s.

On November 17, 1970, the Russian *Luna 17* spacecraft landed the first roving remote-controlled robot, *Lunakhod,* on the Moon. It had a mass of about one ton and was designed to operate for ninety days, controlled by a five-person team using the "move-and-wait" technique. The *Lunakhod* explored the Sea of Rains for eleven months and was heralded as one of the greatest successes of the Soviet space program.

The first *Apollo* lunar rover was flown in 1971 on *Apollo 15.* Two additional lunar rovers went to the Moon with *Apollo 16* and *17.* Each rover weighed about 450 pounds and could carry over 1,100 pounds of passengers and payload. The vehicle was developed in just seventeen months from contract start to delivery—a task made easier because it was designed to be operated by an onboard driver, no computer navigation required. The interplanetary land-speed record of 10.6 mph set by Gene Cernan on the *Apollo 17* lunar rover still stands.

**Astronaut David Scott working at the Lunar Roving Vehicle during the *Apollo 15* mission to the Moon in 1971. COURTESY NASA.**

velopment process: the catastrophic failure of the parachute during a test it was expected to pass without a problem.

"We didn't know if we could recover from it," Adam says. "I estimated that we only had one more chance to get this right. So I had three different parachute designs produced, ranging from ones that used to perform well, but were lower in strength, to ones that were really strong but didn't perform as well."

Under a normal schedule, the usual procedure would be to study the first failure in greater detail, discuss and debate, then choose a single more refined design for the next round of testing. But there was no time for the luxury of methodical advancement.

"We went into test with two of the three designs. Because we *had* to come out of the test with *the* design. There was no other option."

Though no one could know it at the time, MER's multipath strategy worked spectacularly. But just as Pete Theisinger spoke about success in space exploration

Mars *Pathfinder* (shown here) consisted of a small rover with one science instrument and an immobile lander with several. The lander also contained all the radio equipment required to communicate with Earth. For the Mars Exploration Rover missions, *all* the science instruments and communications equipment are carried on a much larger rover, allowing the rover to travel greater distances and take scientific readings at many different sites. COURTESY NASA/JPL.

delivering a "psychic return," so the hard work to achieve that success entails a psychic cost.

Reflecting on his three years of problem solving and hundreds of thousands of miles of travel and those few and far-between weekends with his wife, Adam says, "I don't think any mission will ever be as intense as MER." And for someone who thrives on pressure and challenges and the joy of competition as much as Adam, that thought means something different from the way most others might interpret it. "A lot of things came together to make MER as intense as it was. And one of the things that the Entry, Descent, and Landing team has struggled with is the idea that it may *never again be this good."*

In other words, it was intense, but the team *loved* that intensity, because of the focus it gave them and the ability to do some of their best work ever.

Adam says the members of the MER team aren't the only ones to feel this way. "All the people on Pathfinder talk about it, too: Will it ever be the same for them again? Everyone wonders whether there will just be one of these missions in our lives? Or will there be more than one?"

As someone with knowledge of both missions, Adam—surprisingly—believes the two can't really be compared. "Here's why I believe that my statement of 'There will never be another mission like MER' is different from the Pathfinder statement of 'There will never be another mission like Pathfinder.'

"I think the Pathfinder folks were mostly talking about the fact that they were left alone and they could build things the way they wanted to, and the MER folks are talking a little bit more about the intensity. And it is my sense that I don't know of a team out there that can handle much more intensity than we just did.

"That's kind of like the end of it. I mean, I don't think you can turn up the volume any more and get it done."

And now comes one of Adam's most intriguing insights into a key difference between the pressures faced by the *Pathfinder* and MER teams.

"On MER, we had public pressure. We had *time* pressure. And there was this decision to use the *Pathfinder* plan—the good old *Pathfinder* plan—which had been successful." He smiles ruefully as he adds, "Little did we know."

Know what?

The Mars Exploration Rover parachute being tested in the world's largest wind tunnel at NASA's Ames Research Center. COURTESY NASA/AMES.

## GETTING CARDED

*Rover History, Part Two*

In the mid-1970s, with the success of *Mariner* in 1971 and the development of the Viking mission, Mars exploration came to the forefront of planetary research. But because the round-trip communication delay between Earth and the Red Planet can range from six to forty-five minutes, the "move-and-wait" strategy for controlling rovers on the Moon would not work for rovers on Mars.

Thus, from 1975 through 1980, JPL and the Rensselaer Polytechnic Institute (RPI) focused on developing the technology needed for a vehicle to detect and avoid hazards using onboard sensing and processing—that is, a rover with enough computer "brains" to recognize the edge of a cliff or an impassable boulder and avoid it. Human operators on Earth would be in a "supervisory" role, choosing target locations and planning the proper sequences in which the rover would drive, take pictures, and use its scientific instruments.

These first significant efforts to create a mobile vehicle that could drive itself showed how difficult it was to build a competent autonomous robot. Even connected to the most powerful mainframe computers then available, the first robotic rover could barely navigate as fast as it would under Earth-based "move-and-wait" control.

While the United States was investigating various designs, such as a *"Viking* on Wheels" concept, the Soviet Mars rover *Marsokhod* was being designed for high mobility. Batteries were placed in its wheels to help achieve a low center of gravity, making the vehicle less likely to tip over in uneven terrain. Also included was an "inchworm" mode, which changed the distance between axles when climbing in very soft material. Like NASA's original *Surveyor* Lunar Rover

Vehicle intended for the Moon, the *Marsokhod* never touched the red dust of Mars.

Given the difficulties of making a rover "smart" enough to control its own activities, a technique called Computer Aided Remote Driving (CARD) was developed in the early 1980s. This new technique permitted larger distances to be covered in reasonable safety with commands given in just a single round-trip communication cycle.

Using the CARD system, the rover transmits a stereo image pair—taken by two cameras—showing the terrain ahead of it. Back on Earth, human operators wearing special goggles to help create the illusion of depth study those images on a large computer display. Using the image as a guide, the operators mark key "waypoints" between the rover's current position, and its intended destination. The waypoints are picked to avoid hazards or obstacles. When those waypoints are transmitted back to the rover, the rover uses its own computer to translate the commands into wheel drive and steering to move

from one waypoint to another, in a sense, "connecting the dots."

No more than one or two command cycles per Martian day is practical for a Mars rover. This is because of the time it takes to transmit information from the rover to Earth, the time it takes operators and scientists to plan the next move, and the time it takes for the new commands to be verified, translated into computer-readable format, and sent to the rover.

Recognizing the limitations of CARD, and with the new technology possibilities of the late-1980s, the JPL rover research program next sought to develop "long-range" autonomous navigation. Advances in microelectronics now made it possible to incorporate onboard computing capability that was greater than the remotely connected mainframe computers used in the 1970s. In 1990, the first test vehicle to use these new advances, called Robby (for its likeness to another famous robot), was able to drive itself 330 feet in about four hours. This ca-

**The Russian rover *Marsokhod*.** Note the unique wheel design. COURTESY THE PLANETARY SOCIETY.

pability came close to matching the requirements for the Mars Rover Sample Return (MRSR) mission then under consideration. However, preliminary studies for MRSR concluded that such a mission could cost $10 billion or more, which was not considered practical at the time.

With the decision not to proceed with the large rover MRSR mission, planning then turned toward landing multiple small-instrument packages on Mars. Microrovers, with a mass of around twenty pounds or so, had been first proposed in late 1986. These small rovers were considered a good match with the "small instrument package" idea. Since the operating range for the initial microrovers was thought to be very small and

the computers needed to be small, CARD was considered the only method to use for controlling them.

Each model in the first set of microrover test bed vehicles was called *Rocky*. The name came from their unique mobility system design (invented by Don Bickler of JPL), called a rocker-bogey suspension. This "unsprung" suspension provides exceptional mobility, allowing the vehicle to travel over uneven terrain and climb rocks larger than its wheel diameter very efficiently and safely.

Because it would be far too time-consuming—and possibly fatal—to have the rover stop and radio back to Earth for instructions each time it faced an unexpected

**JPL's "Robby" resembled an entry in a monster truck rally more than a Mars rover. It was thirteen feet long, six and a half feet wide, and weighed more than twenty-six hundred pounds. COURTESY JPL.**

obstacle, a new approach to autonomous driving became necessary.

This new approach, called "behavior control," was developed (initially at MIT) to give the microrover the ability to respond to unexpected conditions, to stop before doing something that could be fatal, like driving off an unseen cliff, using responses that were loosely based on insect behavior!

Adam admits this is a point of contention between him and his colleagues on Pathfinder. But "I got up close and personal with the Entry, Descent, and Landing system on *Pathfinder*. And it is my studied opinion that *Pathfinder* had a nice, healthy dollop of luck that made it happen."

With not a little pride, Adam continues on to describe MER. "I don't believe we had luck. First, we did it twice, which meant the odds of two unusual things happening twice is less. And second, we worked it harder. We worked it harder than they did.

"We had more money than they did. We could spend more money on it. We could do more tests. But we nailed it down tighter than they did."

Adam in no way means to diminish the accomplishments of *Pathfinder*. "I mean, we were standing on the shoulders of giants there." But to redesign the basic *Pathfinder* architecture to make it suitable for the heavier MER mission, Adam explains that the MER team spent a great deal of time analyzing the assumptions of *Pathfinder*'s success, and to his surprise, a couple of those assumptions proved false.

As an example, he talks about what happened when the MER team tried to adapt *Pathfinder*'s parachute design.

"We were doing some testing, and we designed our tests based on *Pathfinder*'s successful designs. But we saw we weren't getting the performance we needed to get, and we said, 'What the hell has happened here?'

"We looked and looked and looked and found, lo and behold, *Pathfinder* had *not* done the test that they thought they had done at all! They were lucky that they were lighter and they were looking for less performance. Because, since they were looking for less performance, they got by on having what was essentially an undertested parachute."

As an interesting side note, the undertesting of the *Pathfinder* parachute came about due to limitations in how the test could be performed, and by not recognizing that the peak loading did not come exactly when it should have. However, this potential underdesign was offset by an overestimation of the drag coefficient, so that, in this case at least, two wrongs ended up making a right.

On MER, Adam says, "We tested our airbags fifty-six times. On the flight design, we dropped twenty-some-odd drops. But *Pathfinder* had a total of sixteen drops over the *whole* development cycle, and only two drops on the flight design.

"So the point is that *Pathfinder* was a grand experiment. It asked the question, Can you do that kind of spacecraft development? And the answer is, Yes. But I think the losses of the Mars Polar Lander and the Mars Climate Orbiter gave another answer: we're [NASA] no longer willing to accept the *risk* associated with guerilla spacecraft development. So MER came back with another challenge: take that kind of landing system and make it bulletproof. That was a much taller challenge than anyone anticipated in the beginning, because memory is always kind, and *Pathfinder*'s success was remembered as just that: success."

In truth, there were so many ways for a Mars lander mission to have ended in failure that no one can argue that luck didn't play at least some part in its success. Somewhere in the *Pathfinder* landing ellipse, on much rockier terrain than MER faced, there must have been a rock sharp enough and big enough to have torn apart

Mars Exploration Rover airbags being tested for resistance to rock impacts in a very large vacuum chamber facility at NASA's Plum Brook Station in Ohio. One of the early uses of Plum Brook was as a testing facility in the late 1960s for equipment designed to be used in lunar outposts. COURTESY NASA/GLENN RESEARCH CENTER.

the airbags on the first bounce. A big gust of wind or the passage of a Martian dust devil could have changed the angle of the backshell rockets at just the wrong moment.

A bit of bad luck might have had the *Viking 1* lander trying to land on that rock—dubbed Big Joe—only thirty feet away that was the same size as itself. And when Neil Armstrong overshot the *Apollo 11* landing site by four miles, had he not spotted a clear patch among the rocks of the Sea of Tranquility with less than a minute of fuel remaining, we might all remember the flight of *Apollo 12* as the first to land humans on the Moon. Looking back even further into aeronautic history, what pilot today would dare risk an Atlantic crossing in so flimsy and untested a craft as Lindbergh's *Spirit of St. Louis*?

Luck—good or bad—will always play a part in exploration. And as navigator Pieter Kallemeyn said, "Every spacecraft that we launch has benefited from the lessons learned from the previous one."

*Pathfinder* did have good luck, but because of its success, the MER team was able to build on *Pathfinder*'s legacy and create a more robust spacecraft, less susceptible to unforeseen conditions. No doubt as the EDL teams for the next Mars lander missions—the 2007 Mars *Phoenix* and the 2009 Mars Science Lab—examine the

## WHEELS ON MARS

*Sojourner and the Mars Exploration Rovers*

The success of the Mars Microrover program led to the adoption of a plan to incorporate a microrover with roughly the same size and control approach as *Rocky* into the Mars Pathfinder mission.

The *Sojourner* rover weighed in at about twenty-three pounds, with another nine pounds of supporting equipment (e.g., radio modem) on the lander. It employed hazard-avoidance sensors which were based on small CCD imagers (used in digital cameras), and laser-stripe projectors, which allowed a wide range of terrain hazards to be detected before the vehicle could be damaged or incapacitated by them.

*Sojourner* operated extremely well on Mars (in fact, better than it ever did in its simulated surrounding on Earth!). The CARD control and behaviors worked well together, allowing the rover's single science instrument, an Alpha Proton X-Ray Spectrometer, to access numerous rock and soil sites. The behavior control kept *Sojourner* out of trouble when ground operators directed her to places they couldn't quite see or to which they misestimated the distance.

The *Pathfinder* lander reached Mars on

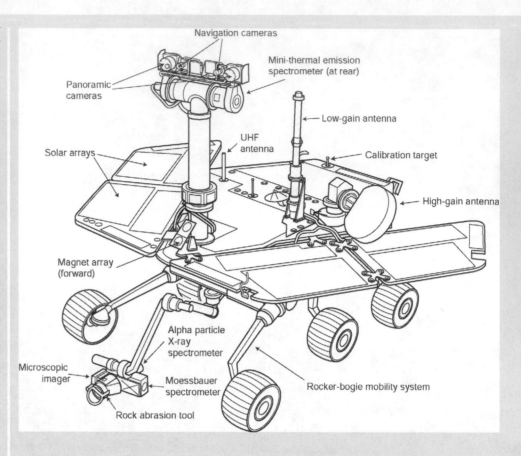

July 4, 1997, and the next day *Sojourner* became the first wheeled vehicle to explore Mars.

The Mars Exploration Rovers were conceived in 2000 as a rebuild of Mars *Pathfinder* with a new rover design that inte-

**The 2004 Mars Exploration Rover. COURTESY NASA/JPL.**

grated all the operational electronics into the mobile vehicle—, *i.e.,* no lander base station. This mission evolved into a twin rover

results of MER in more detail, their findings will result in even better test procedures and "bulletproofing." And a conclusion that MER was lucky, too.

Which leads to an inevitable question for Adam: What will be MER's legacy?

For once, Adam's not certain.

On the one hand, MER's legacy might not be advantageous to new missions. "The people who worked on MER are now very risk-averse, and they're somewhat cost-insensitive." But, Adam adds, "Those of us who have gone through MER, we think that we've had an excellent education."

Another factor that played a part in MER's success, and which might not be applicable to future missions, has to do with a condition it shared with *Pathfinder:* the sense among its team members that this was a once-in-a-lifetime event, which made it easier to justify extreme and highly stressful effort.

mission with an almost completely new design. The first rover, *Spirit*, landed on Mars on January 4, 2004. The second rover, *Op-* *portunity*, followed three weeks later, landing on January 25.

**A JPL family portrait. From left to right:** *Spirit*, **the** *Sojourner* **flight-spare rover**, *Marie Curie*, **and** *Opportunity*. **COURTESY NASA/JPL.**

Specifically, Adam says, "We were able to do MER for less than what its actual cost should have been because of the staggering amount of unaccounted overtime that made *Pathfinder*'s unaccounted overtime—which is famous and regaled—look like a mole hill compared to a mountain.

"We're talking *years* of sixty-plus-hour weeks. We're talking for some folks a year of *eighty*-plus-hour weeks. Just *crazy* amounts of work.

"You know, when you haven't had a weekend for a few months, when you've been working twelve-hour days for a few months, that's why the intensity dial doesn't go any higher than that. Because any higher than that, the wheels are going to come off the cart. Because you can't push people harder than we were pushed."

With the next lander mission he's working on—the Mars Science Laboratory—

This image from the *Viking 1* lander shows the notorious rock named Big Joe. If the lander had come down just a few yards over from its actual landing site, and one of its legs had made contact with Big Joe, chances are the lander would have tipped over and never have been able to return any information to Earth. Though luck has always played a part in every successful landing on Mars, when NASA's next Mars rover, the Mars Science Lab, lands in 2010, its landing site selection will have been made based on detailed hazard maps created by high-resolution orbital images from the Mars Reconnaissance Orbiter. COURTESY NASA.

not scheduled for launch until 2009, Adam is confident the workload of the MSL team will not reach the levels it reached for MER and *Pathfinder*.

"MER shows you can almost come close to overdoing it. We had two *stunning* successes, but if we'd had two failures—which was not out of the question—we would certainly be thinking that the appropriate amount of resources to spend on the problem was a little bit more! [In fact, an independent assessment by outside experts was that the degree of risk undertaken by MER was high primarily due to the lack of an adequate formulation phase, the short schedule, and the complexity.]

"The trick is this—a simple truism: for a spaceflight system to be successful,

## IN MEMORIAM

*The Monuments of Mars*

When the *Spirit* rover successfully landed on Mars, NASA continued a solemn tradition it began in 1982 by naming the *Viking I* lander the Thomas Mutch Memorial Station, to honor the leader of the *Viking* imaging team who died in an accident while mountain climbing. The National Air and Space Museum in Washington, D.C., has a plaque designed to be attached to the Martian lander when it becomes possible for astronauts to visit it in person.

In 1997 NASA honored Carl Sagan by naming the *Pathfinder* lander the Carl Sagan Memorial Station. And in 2001 the *Viking 2* lander was named after Gerald Soffen, the Viking project scientist.

On January 6, 2004, NASA Administrator Sean O'Keefe named the *Spirit* landing site the *Columbia* Memorial Station, in honor of the seven astronauts who died when the Space Shuttle *Columbia* was lost on reentry on February 1, 2003.

A special commemorative plaque was attached to the rover's high-gain antenna before flight and will remain as a fitting memorial on Mars.

On January 27, 2004, to mark the thirty-seventh anniversary of the flash fire that claimed the lives of three *Apollo* astronauts during a launch pad test of their spacecraft, NASA memorialized the *Apollo 1* astronauts by naming three of the large hills seen to the east of the *Spirit* landing site after them.

The next day, January 28, O'Keefe also named the *Opportunity* landing site the *Challenger* Memorial Station, in honor of the six astronauts who died when the Space Shuttle *Challenger* suffered an in-flight breakup during launch, exactly eighteen years earlier.

The plaque commemorating the astronauts who perished when the Space Shuttle *Columbia* was lost on reentry, February 1, 2003. The plaque is attached to the back of the Mars Exploration Rover *Spirit*'s high-gain antenna. COURTESY NASA/JPL.

*hundreds* of people have to do their jobs right. And if any one of *several hundreds* of people screws up, game over."

Perhaps the success of MER is too recent for a definitive analysis of its legacy to be made. The Mars Science Lab mission will require new technology and a new landing system architecture—MER pushed the *Pathfinder* airbag system to its limit. But it has a much more reasonable six-year development schedule, so the heroic efforts associated with *Pathfinder* and the Mars Exploration Rovers should not be necessary.

Adam Steltzner, front and center, leading "a rowdy band of hooligans" to the late-night press conference on *Opportunity*'s landing day.

If so, the development of Mars Science Lab's hardware and software will proceed at a more deliberate and less cost-uncertain pace. But the third component of the mission—the people—might find something missing in that case.

Adam recalls the night of January 24, when the tension of three years finally was over. It left him with one last lesson: the human legacy of MER.

"The trick," he says wistfully, "is to try not to be disappointed. We were all very crushed in the late, late night of *Opportunity*'s landing day after we had this *incredible* success. We were so pumped that we marched into the press conference as a rowdy band of hooligans, barged our way through the doors, and they told us that we couldn't come in because it was too crowded. So we actually sat there later on, and many of us were a little teary-eyed realizing that, you know . . . it's over.

"So the trick is to . . . to adjust to that."

Fortunately, one way to adjust to the end of one mission, is to look ahead, and get ready for what comes next.

It's time to look into the future.

# CHAPTER SEVEN

# "The Point of Going Back"

## The Next Ten Years

*And all of these tales are told in an attempt to give Mars life, or to bring it to life. Because we are still those animals who survived the Ice Age, and looked up at the night sky in wonder, and told stories. And Mars has never ceased to be what it was to us from our very beginning—a great sign, a great symbol, a great power.*

*And so we came here. It had been a power; now it became a place.*

—KIM STANLEY ROBINSON, *RED MARS*, 1993

## THE CONTINUING MISSION

There is only one thing scientists like better than discovering the answer to a question, and that's discovering a new question.

In that regard, if for no other reason, Mars has been a scientific bonanza.

The Mars Exploration Rovers set out to "follow the water," and they did. *Opportunity* landed on what appears to have been the shore of an ancient Martian lake or ocean. *Spirit* landed in a huge basin that looked from orbit like it must have held water, and after a long drive to the "Columbia Hills," has discovered rocks that could have been formed only by the long-term presence of underground water.

But despite these discoveries, the need to follow the water isn't over, because now a hundred new questions clamor to be answered. Where did the water of Mars come from? How long was it there? Where did it go? Does liquid water still percolate below the surface?

The answers are important because, as Firouz Naderi, JPL's manager of Mars Exploration, has said, "Water is a necessary condition for life." For all the many

reasons there are for going to Mars, there is none more important than to find out if life ever arose there, and perhaps still thrives.

Of course, there is a small but vocal community of Mars devotees who believe that the question of life on Mars has already been answered—by NASA's own evidence, no less. And that brings us to this chapter's excursion to another intersection of science and science fiction on Mars. No novel or movie this time, but a real-life conspiracy! Of sorts . . .

## RAIDERS OF CYDONIA

*Is* there a Face on Mars?

The answer to this question is the same as the answer to: Are there UFOs?

The answer to both questions is: Of course!

The skies of Earth are *full* of objects that fly and are unidentified. Whether any of them happen to be alien spacecraft, time travelers, or creatures from other dimensions are conclusions for which no convincing evidence has ever been produced. A true skeptic has to keep an open mind and accept that in such a vast universe, such things *might* be possible. But that same skeptic is also content to wait for incontrovertible proof before accepting the existence of phenomena and/or creatures that could require a fundamental redrawing of the boundaries of our current scientific knowledge and understanding of reality.

Fortunately, in the case of the celebrated face, that incontrovertible proof is now available, and has been for several years.

But first, a recap.

On its thirty-fifth orbit of Mars in July 1976, as part of its mapping mission, the *Viking 1* orbiter took an image of an area on Mars called Cydonia. (The Latin name was adopted by the International Astronomical Union in 1958, following the naming conventions for Mars proposed by astronomer Giovanni Schiaparelli. It was the

The rover *Opportunity* made "following the water" very easy when it landed in a small crater with these exposed sedimentary rocks, later found to have been formed in standing water. COURTESY NASA/JPL.

(OPPOSITE, BELOW) The story of water on Mars was continued at *Spirit*'s landing site. The rock seen dead ahead, named Mazatzal, became the target of the rover's RAT—Rock Abrasion Tool. The RAT took three hours and forty-five minutes to carefully grind away the outer layers of the rock. Microscopic close-ups of the grind revealed details that tell geologists that water was also involved in the formation of this rock, though it was likely groundwater, and not standing water as at *Opportunity*'s landing site. Credit for navcam image: COURTESY NASA/JPL. Credit for inset image: COURTESY NASA/JPL/UNITED STATES GEOLOGICAL SURVEY/CORNELL.

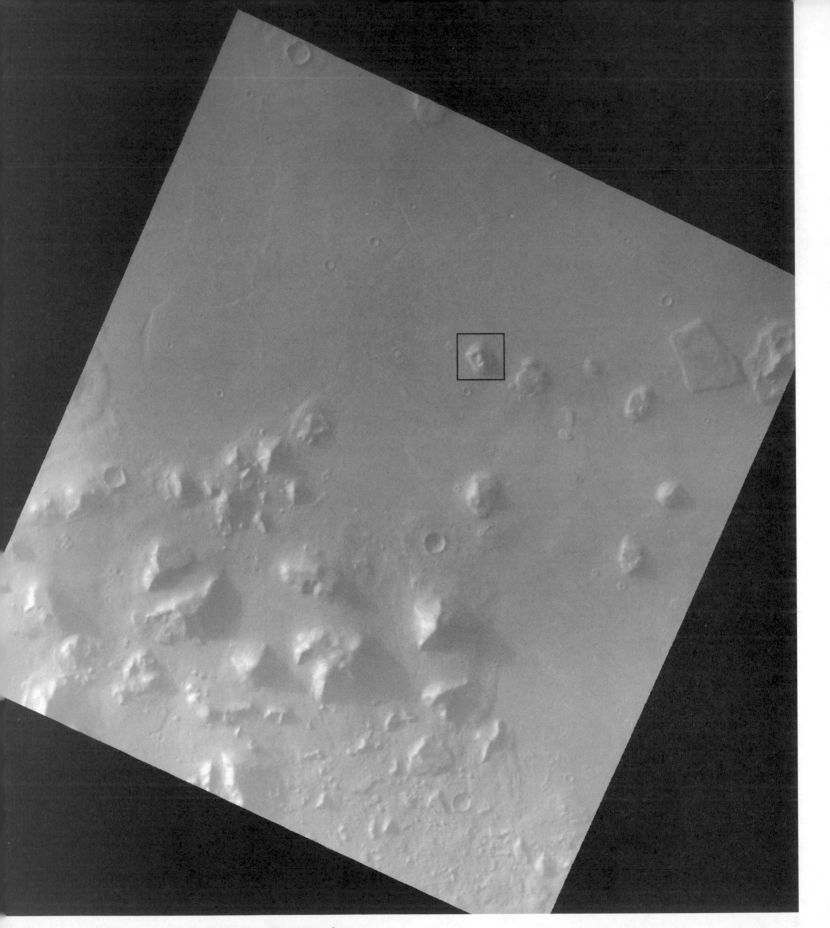

The infamous "face" (in box) that launched a thousand conspiracy
stories. This is the first image of the so-called Face on Mars, taken
by the *Viking 1* orbiter on July 25, 1976. The facelike mesa is about
one mile long. COURTESY NASA/JPL.

The highest-resolution image of the "face" obtained to date, clearly showing it to
be an eroded mesa. COURTESY NASA/JPL/MALIN SPACE SCIENCE SYSTEMS.

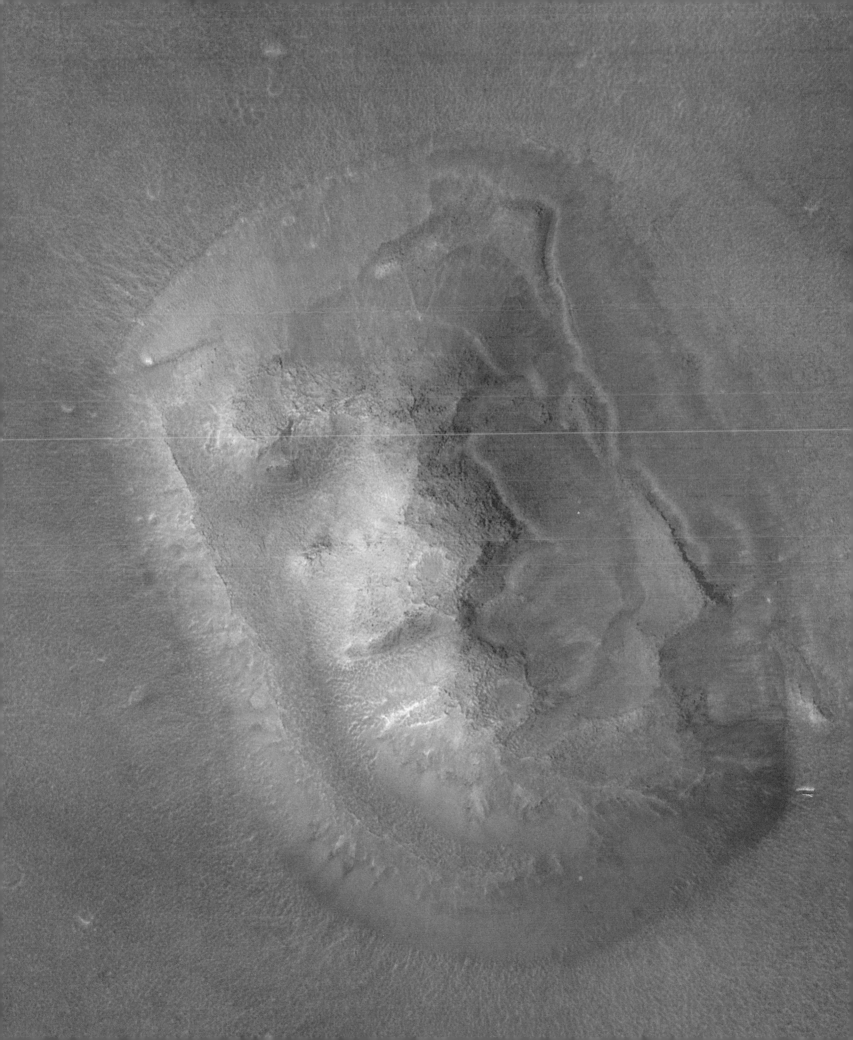

name of a city on Crete.) A few days later, that image—35A72, for the seventy-second image taken by orbiter A on the thirty-fifth orbit—was printed out on Earth and examined by the team looking for a safe landing site for the *Viking 2* lander.

*Viking* imaging team member Toby Owen is credited with being the first to see the printed frame, and immediately noted the image of a human face, albeit about a mile long. It was such an unusual sight that the imaging team made a point of showing the image at a press conference. At the time, there was little question in anyone's mind that the "face" was nothing more than a trick of light and shadow on a completely natural geological feature. When the feature was imaged again under different lighting conditions, no doubt the face would no longer be there, because the light and shadow angles would be different.

But then, Cydonia *was* imaged again, thirty-five days later, on frame number 70A13. Intriguingly, though the angle of lighting had changed by about twenty degrees, the feature was still recognizable as a face.

Of such things are legends born.

The vast majority of scientists studying Mars continued to see only a random pattern that appeared to have facelike characteristics—a pattern, after all, which the human mind has been genetically programmed to recognize. At the same time, however, a quite different opinion began to gain followers. A group of enthusiasts involved in the fringe areas of science proclaimed the Face on Mars—which had now acquired capital letters—to be an artificial structure, built either by Martians or by alien visitors to Mars.

Soon other structures were "discovered" in the *Viking* orbiter images: "pyramids" and vast "cities" arranged with the Face in complex mathematical relationships which were compared to hidden patterns in the arrangement of Earth's own Egyptian pyramids.

To the believers, the evidence was overwhelming and led to many more questions. Why was the Face built? What were the Martians (or other aliens) trying to tell us? And why was NASA so insistent in denying that the Face was artificial? What did NASA—and by default, the U.S. government—know about aliens, and when did they know it?

If this all sounds familiar, it should. The groups that sprang up around the study of Cydonia were little different from those that proclaimed UFOs were alien craft. Common to all is the belief that *someone* in authority knows the truth, and that that truth is being withheld for nefarious reasons.

But UFOs, by definition, tend to be a fleeting phenomenon. After all, if an unidentified flying object hangs around within camera range long enough, chances are it's going to be identified. The sightings that feed the myth are the ones that got away.

The Face on Mars, however, was something unique in the realm of pseudo science—the possibility that it was an artificial structure was what scientists call "a testable hypothesis." That is, further study could definitively prove whether or not it was artificial. Its location on Mars was known, and all it would take was a return visit with a new and improved camera capable of taking more-detailed images.

In the late '70s, NASA was definitely stodgy about discussing the more science-fictionish possibilities of the ongoing exploration of space. (Though many NASA employees have the quick imaginations and sense of humor common to those on the cutting edge of science and technology, NASA's unofficial policy was to keep any sense of frivolity swept under the bureaucratic carpet. But despite NASA's desire to ignore the topic, interest in the Cydonia region did not go away, and NASA serves the public. Reluctantly, the agency announced that when the Mars Observer orbiter flew over the region on its 1993 mission, the Face would once again be imaged, this time in greater detail.

But the Mars Observer disappeared on August 21, 1993, only three days before its arrival at Mars. For believers in the Face on Mars, it was an *X-Files* episode come to life. NASA engineers concluded the most likely reason for the failure was a flaw affecting the probe's propulsion system. But Face on Mars believers smelled conspiracy. Some even maintained that the Mars Observer didn't disappear at all, but that NASA spread that story in order to keep its detailed images of alien activities on Mars top secret.

Four years later, when the Mars Global Surveyor (MGS)—carrying the flight-spare camera from the lost Mars Observer—arrived at Mars, once again NASA announced that Cydonia would be imaged, whenever the region came within range.

Then, as the Face on Mars believers anxiously awaited the images that would prove their case, the MGS was discovered to have a problem with one of its solar panel hinges. While the probe had been designed to achieve its final orbit at Mars through a series of aerobraking maneuvers that required it to skim through the upper Martian atmosphere, NASA engineers decided those maneuvers could now result in the loss of the solar panel. So a more cautious and much slower approach was taken, and the MGS mapping mission that was to have begun in September 1997 was instead delayed to March 1999.

Inevitably, the conspiracy rumors grew again. A small but vocal group disputed NASA's account of events. Clearly the agency would stoop to any means to prevent Cydonia from being imaged in detail. Something was being held back, but what?

The truth was finally captured on April 5, 1998, and revealed to the world when the Mars Global Surveyor image was received the next day. Though the full MGS mapping mission could not start until the probe's orbit had been finalized, one of its passes took it over Cydonia, and true to NASA's word the entire region was imaged as planned, this time at a resolution ten times higher than the original *Viking* frames.

Truth be told, there probably were some disappointed people at NASA that day. If the new photos of the Face on Mars had revealed clear evidence that it was an artificial construction, NASA's funding for Mars exploration would have skyrocketed, making the budget for Project Apollo look like lunch money, and today there might have been teams of astronauts staffing permanent science outposts at Cydonia.

But, alas, the new photos revealed exactly what Mars experts expected: the Face on Mars was an eroded mesa, nothing more.

If the *Viking* orbiters of the '70s had had the ability to take similar high-resolution images, so an immediate closer look could have followed the original im-

## THE PHOTOS NASA DOESN'T WANT YOU TO SEE!

*Can NASA Censor the News from Mars?*

As far back as 1952, when George Adamski took a photograph of a chicken brooder lamp and claimed it was a flying saucer from Venus, space exploration and questionable photographs have had a close, if uncomfortable connection.

Over the years, some conspiracy-minded people have gone so far as to say that *all* the *Apollo* photographs showing astronauts on the Moon were faked by NASA. The idea of faking a landing on Mars was even the premise of the 1978 film *Capricorn One.* And for every conspiracy buff who argues that the *Apollo* photos are hoaxes, there's another who claims the very same photos show evidence of mysterious alien structures on the Moon.

Given this rich history of imaginative interpretation of NASA imagery, it comes as no surprise that within hours of the first photographs from the *Pathfinder* and Mars Exploration Rover missions being posted on the Internet, extraordinary new claims were made about what the photos really showed.

Some people believe that some images captured alien spacecraft in the skies of Mars, hovering near the *Pathfinder* landing site. If that's true, then the aliens are extremely knowledgeable about Earth technology, because they have disguised their craft to look exactly like "compression artifacts"—those patterns that appear when digital photographs are compressed to take up less computer memory.

Apparently, Martians fell on hard times; between the *Pathfinder* landing of 1997 and the MER landings of 2004, only scattered alien debris was found.

Instead of following the procedures established by earlier NASA projects, which allow lead science teams an exclusive period to review scientific data returned from spacecraft, the *Pathfinder* team put considerable effort into publicly releasing all their imagery from Mars as fast as Eric Dejong and the digital magicians of the Multimission Image Processing Lab could capture and produce them, sometimes within minutes, of being received at the Jet Propulsion Lab. The same procedure was followed for both MER

missions, but even faster and better. This meant that raw pictures were posted on the Internet, without benefit of any enhancing or smoothing programs, clearing the way for other people to "enhance" the photos as they saw fit.

For example, some doctored images show different versions of Twin Peaks, the rounded hills near the *Pathfinder* landing site. While the original, raw version shows only a rocky hill, some people claim the enhanced versions show a ruined wall, perfectly round rocks, and twisted metal structures. In fact, given the number of alien structures claimed to surround the *Pathfinder* landing, one of the most remarkable feats of the mission appears to have been the way the lander managed to avoid crashing into any of them.

Of course, just as with the *Apollo* photographs, there are those people who firmly believe the *Pathfinder* photographs can't possibly show alien structures on Mars, because the spacecraft never reached Mars in the first place. Instead, these people claim, all the photographs received from the lander were faked.

In the end, the only conclusion that can be drawn from the existence of those diametrically opposed viewpoints is that among some people, at least, there seems to be an almost perverse need to believe that NASA has regularly achieved one of its greatest goals—"To discover life out there"—only to try to keep that stunning accomplishment secret, while at the same time inadvertently and ineptly allowing the evidence to slip out.

Well, you can't have it both ways.

There is not one person working for NASA who would not shout for joy if a Mars rover sent back a photograph of an unambiguous fossil or the clear chemical signal of a biologi-

Because NASA's policy is to release raw images from Mars as soon as possible after they are received on Earth, hundreds of artists delight in using photo-manipulation programs to create their own "unique" versions of those images. In truth, there is nothing NASA would like better than to actually find something like this on Mars, because such a discovery would lead to a significant acceleration of current plans to explore the Red Planet. ALTERED IMAGE BASED ON ORIGINAL IMAGE COURTESY OF NASA.

cal process. Not only would such a discovery fulfill one of NASA's primary missions, it would fuel an unprecedented new era of Mars exploration and inevitably lead to welcome increases in NASA's budget.

But even if there were some reason why literally thousands of independent minded scientists and engineers who would like nothing better than to find evidence of extraterrestrial life decided to keep a discovery a secret, could NASA actually pull it off?

In a word, no.

NASA does not consider itself a media production organization. As a matter of policy and practicality, the agency distributes a raw feed and lets the scientific community,

the media, and in this Internet era, the public, digest and distribute it as it happens. As anyone can tell who has watched the live broadcasts from the Jet Propulsion Lab showing deliriously happy people cheering as the first pictures arrive from Mars, if one of those photos had shown a Martian critter, millions of other people would see it at the same time as NASA personnel.

Indeed, during the Mars *Pathfinder* mission, the lumpy rock called Barnacle Bill was given its name because in the very first photographs of it, the *Pathfinder* science team thought those lumps really might be fossilized barnacles. There was no cover-up, no flurry of encrypted phone calls to the mysterious keep-

The Mars *Pathfinder* rover *Sojourner* making its way past the rock (on left) named Barnacle Bill. The rock earned its name because the first images of its surface texture suggested it might actually be covered by fossil barnacles—an exciting possibility that did not survive closer examination. COURTESY NASA/JPL.

ers of the Mars conspiracy. Instead, the team called the rock as they saw it, and, like good scientists, instead of voicing wild speculations, made plans to test the hypothesis with the *Sojourner* rover by driving over and taking a closer look, knowing full well that America, and the world, would be looking over their shoulders the entire time.

NASA wouldn't have it any other way.

ages, no doubt the Face would have been only a small footnote in the annals of Mars geography.

But in the twenty-one years that passed between the first images taken by *Viking* and the definitive ones taken by Mars Global Surveyor, the Face spawned scores of books and videos and legions of researchers who wove those two original *Viking* frames into vast conspiracies and outlandish "scientific" theories, many of them contradictory, and several providing a decent living for their promoters. The Face was no longer just an interesting anomalous image, subject to legitimate scientific study. It had become an industry.

A few patient scientists have taken the time to refute the claims of the Face on Mars believers, who now maintain that NASA and/or its contractors continue to deliberately manipulate MGS images to hide the truth. But for those true believers, NASA photographs are no longer necessary. The Face on Mars has passed from any possibility of being a testable hypothesis to becoming a matter of faith, and science has no role in that debate.

As a last word on the subject, it is telling that the primary promoters of the Face as an artificial structure also analyzed images returned from the Mars Exploration Rover *Spirit*, identifying manufactured alien artifacts scattered around the rover's landing site as if it had set down in a Martian junkyard. Yet, by an odd coincidence, all of the so-called artifacts were somehow positioned just outside the rover camera's limits of focus and resolution, making them as blurry and as inconclusive for scientific examination as those first frames of the face returned from *Viking*.

There are legitimate mysteries on Mars that once solved, may forever change the way we think of ourselves and of our place in the universe.

But the Face on Mars isn't one of them. Those who cling to specious allegations put forth by its promoters only waste the time of the real researchers who dedicate their lives to the demanding work of evidence-based science and discovery.

Now it's time to join some of those people who will take us on our next steps on our journey to Mars—steps that might finally bring us, if you'll pardon the expression, face to face with *real* Martians.

## FORGET ABOUT VELCRO AND TANG

In June 2004, the President's Commission on Implementation of United States Space Exploration Policy delivered a report that will serve as a roadmap for NASA's next several decades in space.

In 1958, the year NASA was formed, the agency's goals were basic: to learn how to send machines and humans into space and have them function there. The explosion of scientific knowledge and technological achievements that emerged from the accomplishment of those goals has changed the United States, and most of the world, into a space-based civilization. It used to be that NASA expended considerable public relations effort trying to justify its use of taxpayer dollars by endlessly promoting the spin-offs that came from its technological breakthroughs.

Detractors disparaged those efforts by making jokes about Velcro and Tang (neither of which was invented by NASA, though both products were subsequently used on space missions).

Today, anyone who questions the benefits of an ongoing and aggressive space program just isn't paying attention. Global communications, weather prediction, computer modeling, cell phones, robotics, artificial intelligence . . . To ask about the societal benefits of space exploration is a bit like asking about the societal benefits of medical research.

The reality is we live in a technological age created in part by NASA. And the continuing exploration of space, now focused on even grander goals requiring even more technological brilliance, will transform our world, scientifically *and* philosophically.

According to the 2004 Presidential Commission, the journey NASA will embark on in the decades ahead will fulfill a science agenda organized around these broad themes:

ORIGINS. The beginnings of the universe, our solar system, other planetary systems, and life.

EVOLUTION. How the components of the universe have changed with time, including the physical, chemical, and biological processes that have affected it, and the sequences of major events.

FATE. What the lessons of galactic, stellar, and planetary history tell us about the future and our place in the universe.

It doesn't get bigger than that.

Some of the first steps toward meeting this agenda will be taken on Mars, first by ever more capable machines, and eventually by humans.

No one can say how many steps the journey will require, so NASA's Vision for Space Exploration is open-ended—"one mission, one landing at a time." As new discoveries are made at one step, new strategies will emerge for the next. Looking far into the future, NASA anticipates permanent outposts on the Moon and human expeditions to Mars. But exactly how and when those historic milestones will be accomplished is dependent on the results of a carefully planned series of achievable and affordable precursor missions.

It is some of those missions we'll look at now—part of NASA's plans for the next ten years in the exploration of Mars.

To guide us, we'll rely on one of JPL's most experienced "Mars hands"— Richard Cook, whose first trip to Mars was as mission operations manager for the spectacularly successful Mars *Pathfinder*. Soon after, he experienced the '98 failures personally, as operations phase project manager for the lost Mars '98 missions, then soared back to Mars as the flight systems manager, and then project manager, for the 2004 Mars Exploration Rovers.

Today, Richard is at the forefront of the next major rover mission to Mars,

## WHAT DID THEY KNOW AND WHEN DID THEY KNOW IT?

*The Intriguing Tale of Jonathan Swift and the Moons of Mars*

Not all the mysteries of Mars arise from the latest scientific discoveries. The mystery of Jonathan Swift and the Martian moons goes all the way back to 1726, when Swift published his most famous novel, *Gulliver's Travels.*

In that book, Gulliver visits the island of Laputa, which floats in the air and is home to a group of accomplished astronomers. Swift writes that the Laputans have "discovered two lesser stars, or satellites, which revolve about Mars; whereof the innermost is distant from the center of the primary planet exactly three of its diameters, and the outermost, five; the former revolves in the space of ten hours, and the latter in twenty-one and half . . ."

What's so remarkable about Swift's oddly accurate description of the moons of Mars is that they were discovered by astronomer Asaph Hall in 1877—151 years *after* Swift described them.

Since there were no telescopes powerful enough to see the Martian moons in Swift's time, most people believe he simply made a lucky guess. One thing that might have inspired him to make that guess was the belief in his day of the way the solar system was arranged. Mercury and Venus had no moons. Earth had one, Jupiter four, and Saturn, five. That seemed to indicate that Mars should have two moons, but since they hadn't been detected yet, that meant they had to be small and orbiting close enough to the planet to be obscured by its glare.

Thus, like a true science-fiction writer—though the term would not be used for another two hundred years—

serving under Peter Theisinger as the deputy project manager for the 2009 Mars Science Laboratory.

## RICHARD COOK

"I really, really like doing it when it works like this."

Those were the first words Richard Cook spoke at the postlanding press conference for the *Spirit* rover, January 3, 2004. Unlike some of his colleagues who were reliving the success of Mars *Pathfinder* that night, Richard was also remembering the pain of losing the Mars Polar Lander in December 1999. But NASA and JPL learn their lessons well, and from the '98 failures came the '03 successes.

For all that Richard Cook savored the moment of *Spirit*'s success, to work in the exploration of space had not been his goal as a youth. "I sort of fell into this," he admits.

But his pursuit of at least some type of career in engineering wasn't as unexpected as Adam Steltzner's had been. "Obviously, I sort of knew that I wanted to do math and science," he says. "But when I left high school, I think like a lot of high school kids, I wasn't sure exactly what I wanted to do—there were a lot of possibilities."

Some of the possibilities Richard considered ran counter to Pete Theisinger's youthful assessment of what those drawn to science and math find interesting. "I'm a big history person," Richard says, "so I thought about getting involved in that. And then I really liked economics and stuff like that. So I thought about that, as well."

But engineering was always among those possibilities, too, and by the time Richard entered college, he knew his path was taking him to either that or science.

Like many people at JPL—again, with the exception of Adam Steltzner—science and engineering were part of Richard's early life. His father was a geophysicist for an oil company. "Also, I grew up in Denver, Colorado, so I lived about fifteen minutes from what is now Lockheed Martin Aerospace, and I had a lot of friends whose parents worked there."

Like most teenagers with an interest in science, Richard was "into space." But not to any great degree. "If you'd asked me if I watched the *Viking* landing or was even aware of *Viking*, I probably would have said no."

Richard attended the University of Colorado at Boulder. He started in what he calls "their version of applied physics," which he liked because "It allows you to not decide what you're going to do."

In that program, students majored in physics with an engineering minor. Richard started in aerospace engineering. "I took one class and decided it was terrible. So I decided to become an electrical engineer."

Unfortunately, after a year and a half of those courses, he decided electrical engineering was terrible, too. But he had a decision point fast approaching. "The way the program worked is, you could either decide to set yourself up so you go to grad school in engineering, or set yourself up to go to grad school in physics.

Swift described his literary invention as having all the characteristics that would make it impossible for anyone to say his invention wasn't true.

Mystery solved . . . or is it?

"For a long time I toyed with doing astrophysics and even tried to get jobs at radio observatories. But none of that happened. So, eventually, by the time I got to be a junior, I had decided I had had enough electrical engineering. I went to the chairman of the aerospace department because I figured maybe I'll give aerospace a shot again.

"I went and talked to him and I remember saying, 'I'm sort of interested in what possible things I could do in aerospace.'" Richard points out that aerospace was an important term. "I'd taken a fluids mechanics class the first semester I was a freshman and I hated it. So I didn't like aeronautical engineering."

Richard was quite matter-of-fact in his meeting with the chairman. He didn't like electrical engineering. He didn't like aeronautical engineering. He was "sort of" interested in aerospace engineering. So he asked a perfectly logical question: "Is there other stuff?"

Fortunately, the chairman had the right answer. "He said, 'Well, you're interested in physics, so you must like orbital mechanics.'" The first thing that Richard thought was, What's that? But he replied, "Sure, of course."

Richard quickly found out that orbital mechanics "is basically the dynamics of spacecraft traveling to and around bodies in space. I took a class in that and I found it to be very interesting. So I took another class." Richard had found his calling.

After earning his undergraduate degree, Richard studied aerospace engineering at the University of Texas, with a specialty in orbital mechanics, for which he earned his master's.

But after seeing that some students at Texas spent up to ten years obtaining their PhDs through the university's prestigious Center for Space Research, Richard decided he was ready for the job market. As he explains it with considerable self-insight, "I came to the conclusion that it's possible that I wasn't cut out to get a PhD because I just wasn't interested in knowing anything that well." Then he adds, "And actually, it's sort of funny, but I think the more education you have, the more you realize how little you know."

As it turned out, Richard's decision to enter the job market was perfectly timed, and he received numerous job offers, including one from JPL. "Texas is very well regarded, and there's almost a Texas-to-JPL pipeline of people. JPL is responsible for mission and trajectory design, which is a subset of orbital mechanics in that you're worried about how to get an object from the Earth to somewhere else." The pipeline was still in operation, and Richard was hired by another Texas graduate to work at JPL doing trajectory design.

Considering he got job offers from a number of different potential employers, why did Richard choose JPL? "Mostly because of the work," he explains. "They also paid the most, which is certainly part of it." With a smile, though, he quickly discounts that reason. "But paying the most and coming to California is a double-edged sword. I mean, they have to pay the most because it's expensive here. So, that wasn't really too much of a factor. It was mostly that the work was going to be pretty exciting."

Like many graduates hired by JPL, Richard was offered his choice of two different positions. "One was to work in this overall trajectory section that does very detailed trajectory work for missions that are in development or about to fly. The other was for a section that does what is called advanced projects where you start right at the beginning of a mission to find out what's possible. I chose to get into that area because I'd be in at the beginning. I'd have more of a big picture."

Richard's first assignment was to work on a study for a lunar orbiter mission. "At the time, there were a lot of issues associated with what it would take to do that, including how you got to the Moon, and how stable an orbiter around the Moon would be. So we had to build up a bunch of software and capability to study that particular problem. I spent, probably, two years looking at that kind of a problem."

Keep in mind that this was in 1989, when dozens of spacecraft had been to the Moon and successfully entered orbit. More evidence, if any were needed, that navigation in space is both a science and an art.

In the end, though, as seems to be traditional for the study projects to which new JPL employees are assigned, the lunar orbiter project never proceeded. Richard's next assignment was to work on a mission that was already flying: the *Magellan* Venus orbiter.

"We were looking at new ways to do what's called aerobraking, which is where you use the atmosphere to slow yourself down. We were studying ways to do that, so I got very involved." The challenge of that particular assignment was that the *Magellan* hadn't been designed for aerobraking. But since the mission was reaching its end, the orbiter could be exposed to greater risk for experimental purposes. In the end, Richard says, "It actually worked very well. It was very successful."

Finally, in late 1991 Mars showed up on Richard's radar, and he began studying the possibilities for the Holy Grail of current Mars missions—the Mars Sample Return (MSR). The technical complexity of MSR was overly ambitious for the time, but when that particular study ended, another Mars study appeared—MESUR Pathfinder.

Sound familiar?

Richard began working on what would become the Mars Pathfinder mission in November 1991, long before it *was* Mars Pathfinder. He recalls that one of the others working on the project was Matt Golombek, but it was not until February 1992 that Tony Spear came onboard to be project manager.

Richard had chosen his first position at JPL in order to have a chance to get in on a project at the very beginning, and less than two years later, he had his wish—and it was taking him straight to Mars.


## BRINGING IT ALL BACK HOME

Between 2005 and 2014, there are five launch opportunities for Mars missions, and NASA has current plans for them all.

Two missions are as definite as any space venture can be: their hardware has

## THE ULTIMATE ROAD TRIP

*Mars Science Lab*

The Mars Science Laboratory (MSL) represents a new generation of Mars explorer leading into the next decade of Mars surface missions. The mission science objectives are to understand the "habitability" of Mars—that is, the capacity, in the past, present, and/or future, for Mars to sustain life. This objective will be met through multidisciplinary measurements related to biology, climatology, geology, and geochemistry.

MSL is a long-duration, high-performance rover that is inherently more robust to the unknowns of the Martian environment. By virtue of its onboard capabilities and large payload, it has the potential for making significant new scientific discoveries.

MSL is proposed to launch in late summer or early fall 2009, and to land on Mars in May 2010. In order to be operational at latitudes reaching to the poles at any season, for as much as a full Mars year (approximately two

Earth years), MSL is considering using a radioisotope power supply as was used on Viking.

**A preliminary design for what is planned to be NASA's next Mars rover—the Mars Science Lab. COURTESY NASA/JPL.**

been designed and is currently being built. These missions call for a Mars orbiter and a Mars lander: the 2005 Mars Reconnaissance Orbiter (MRO) and the 2007 Mars *Phoenix* Lander.

The Mars Reconnaissance Orbiter mission will map Mars in unprecedented detail, with a high-resolution camera able to discern details as small as *Pathfinder*'s *Sojourner* rover. Some of the MRO's mission objectives will be to identify landing sites for future missions, to search for underground water, and to serve as a telecommunications link for landers to come. The maps it will help produce will be of such high quality that future landers may be able to target landing sites of the greatest scientific interest with unprecedented precision, possibly even using a concept called "pinpoint landing" based on cruise-missile technology. This approach will not only greatly improve the efficiency of the scientific missions, it will also reduce risk by avoiding landing hazards like large rocks, steep slopes, and small craters.

And while MRO will use radar to look for underground water from orbit (as does the MARSIS experiment on the European Space Agency's Mars Express orbiter), the 2007 *Phoenix* lander will look for "ground truth." That is, in a polar region where near-surface ice has already been detected from space, the *Phoenix* will use its robotic arm to dig down to that ice, and then use its onboard science instru-

## SIZE DOESN'T MATTER

*Mars Scout Missions, 2011*

The Mars Scout program is designed to produce a class of small, relatively low-cost missions that supplement major undertakings such as the billion-dollar-plus Mars Science Lab rover. And where major Mars projects generally originate with NASA Headquarters, Scout missions are termed "fully competed." That is, NASA opened the doors to qualified participants to propose any mission they wish, provided it fits within the limited Scout budget of $325 million and—for the first mission—would be ready for launch by 2007.

For the 2007 Scout mission, NASA selected the Mars *Phoenix* Lander from among four "finalists." NASA Associate Administrator for Space Science Dr. Ed Weiler has called the other three missions "outstanding proposals for exploring Mars on a modest budget, to answer several priority questions about the Red Planet." These missions will likely be proposed for the Scout opportunities scheduled for 2011.

Here's a look at what might be flying to Mars in the future.

**SCIM: Sample Collection for Investigation of Mars**
Principal Investigator: Professor Laurie Leshin, Arizona State University, Tempe.

A very clever Mars Sample Return mission that doesn't require landing on Mars. Instead, a small spacecraft will skim through the upper atmosphere of Mars, capturing airborne particles, then return them to Earth for detailed analysis. NASA believes, "Such samples could provide breakthrough understanding of the chemistry of Mars, its surface, atmosphere, interior evolution, and potential biological activity."

**ARES: Aerial Regional-scale Environmental Survey**
Principal Investigator: Dr. Joel Levine, NASA Langley Research Center

If ARES is selected, it really will be flying, because it calls for a small airplane—specially designed for the thin Martian atmosphere—to unfold from a descent capsule and then travel above the surface of Mars at an altitude of about 3,300 feet and at a speed of about 300 mph propelled by a rocket engine. The airplane's instruments will analyze the near-surface atmospheric chemistry of Mars, which has never been done, and which could eventually help reveal how the climate of Mars has changed.

**MARVEL: Mars Volcanic Emission and Life Scout**
Principal Investigator: Dr. Mark Allen, JPL

This is an orbiter mission, based on the successful Mars *Odyssey,* but with new instruments designed to study the Martian atmosphere for emissions that could be related to active volcanism or microbial activity. The MARVEL orbiter will also be able to observe the behavior of water in the Martian atmosphere over the course of an entire Martian year.

If ARES is chosen as one of the next Mars Scout missions, it will launch into the Martian atmosphere the first winged vehicle to fly on another planet. COURTESY NASA.

ments to examine the ice for indications of, among other things, biology-related chemistry.

For the 2009 launch opportunity, design work is under way for the Mars Science Laboratory (MSL)—a new rover mission carrying a large scientific payload capable of a wide range of measurements of biological compounds on Mars. The MSL will be a rover the size of a Mini Cooper car, capable of traveling across Mars for as much as a full Martian year (687 Earth days). Though MSL has yet to receive its official confirmation to proceed with implementation to launch, its preliminary design phase is proceeding on schedule.

Since the 2009 opportunity is favorable for landers, the possibility remains that a second rover might also be launched to explore a second area on Mars. Because Mars Science Lab might use a radioisotope power supply, its potential landing sites are not limited to equatorial regions to which solar-powered vehicles are generally restricted. That means many more scientifically interesting areas of Mars are open to it.

For 2011, a specific choice has not yet been made. But plans could include two new "Mars Scouts"—limited-scope, relatively low-cost missions designed to supplement NASA's major missions. (The 2007 *Phoenix* Lander will be the first Scout mission—its comparatively low cost made possible because it is based on hardware built and paid for as part of the canceled '01 Mars *Surveyor* mission.)

New Scout candidates for launch in 2011 could contain concepts built on past proposals, involving a fascinating variety of imaginative concepts, including a winged aircraft to observe Mars from an altitude of about three quarters of a mile; a low-cost sample-return mission that will skim through the upper Martian atmosphere to collect dust samples for return to Earth; a torpedo-shaped probe that will melt its way into the polar icecaps; and a set of gliders with six-foot wingspans to explore the Valles Marineris canyon—the six-mile-deep, 2,500-mile-long "Grand Canyon" of Mars.

But it's the 2014 launch opportunity that could mark the beginning of one of the most important, exciting, and complex Mars missions yet—Mars Sample Return (MSR).

The date is not definite since extensive study and development are still required to finalize the best way of achieving the ambitious goal of scooping up Martian rocks and soil and bringing them back to Earth. Engineering and cost issues have already resulted in the cancellation of an MSR mission planned to begin in 1993, and another for 2001. But Richard Cook describes one possible way the multistage mission might unfold in 2014 or beyond.

"The first stage is a rover lander with a Mars Ascent Vehicle (MAV). It will land, the rover goes out, picks up the samples, then brings them back to the MAV. The MAV then launches and goes into orbit of Mars." It would then be picked up by an orbiter and returned to Earth. Sounds pretty simple, doesn't it?

Under this plan, two separate rover/MAV landers might be launched to Mars, resulting in two sets of samples in orbit of Mars. Original ideas about MSR assumed that just collecting some material, any material, would be good enough. However,

following the MER discoveries, scientists are much more interested in getting samples from specific sites where the potential for finding evidence of water-based and even biologic processes is greatest (i.e., Meridiani). In fact, some scientists have made the case that for the MSR mission to be scientifically useful, obtaining samples from up to ten different locations on Mars should be considered.

Richard marvels at the technical challenges such a mission presents. "It's definitely the most complicated interplanetary mission ever done. It's probably one of the most complex missions ever done, period. Just the fact that everything takes place millions of miles from the Earth . . ." With a shake of his head, he recalls that first project he worked on for JPL—the lunar orbiter mission. "Rendezvousing around the Moon is easy compared to Mars because for the Moon, it takes less than two seconds for signals to reach the Earth." At Mars, with signals taking many minutes to be received by ground controllers, the entire rendezvous procedure will have to be fully automatic.

"Plus," Richard continues, "we've never built a rocket to lift off the surface of another planet with an atmosphere. We did for the Moon, but never from anywhere with an atmosphere. *And* we've got to build a rover to go out and get the samples—we've never brought back samples that a human didn't pick up. The Russians did, though," he adds, referring to Russia's *Luna* series of successful lunar-sample return missions in 1970, 1972, and 1976.

Of course, the Mars Science Laboratory will address the issues of sample acquisition and delivery and mission planners have been asked to consider the possibility of caching samples for a future MSR mission. This could involve taking and delivering samples to a container of some kind that would then be made accessible for a future mission to retrieve. This approach requires very precise targeting of the sample return lander to wherever the MSL rovers have left the sample container. But it takes advantage of the sophisticated capabilities of MSL to collect and prescreen samples using its onboard instruments. At the time of this writing a science group is working to develop the requirements for such an augmentation of the MSL mission.

But in addition to the technological challenge of a Mars sample-return mission, there is another major hurdle to overcome. "On top of it all is the fact that we're bringing rocks to the Earth from Mars, and there're a lot of people out there who are a little concerned about that."

"A little concerned" is the understatement of the decade. *If* there is life on Mars, and Martian microbes hitch a ride to Earth in the collected samples, then the prospect of Earth's biosphere being contaminated by an alien life-form is no longer just within the realm of science fiction—it's a definite possibility. Even if life on Earth and life on Mars share a common ancestry, as some scientists believe might be possible, the chance of accidental contamination is potentially hazardous, especially if the reentry capsule carrying the samples through the Earth's atmosphere should break up in the air or shatter as a result of a hard impact.

Tom Rivellini, who has also worked on different MSR studies, explains how NASA approaches that concern.

"Landing on Mars, returning to orbit, robotic orbital rendezvous—on a relative

(OPPOSITE PAGE) This concept illustration shows a Mars Ascent Vehicle (MAV) launching from Mars with its cargo of rocks and soil. A Mars Sample Return (MSR) mission could consist of several missions working together: several rovers to gather samples and deliver them to a MAV; one or two MAVs to take the samples into Mars orbit; and at least one Earth Return Vehicle (ERV) to autonomously dock with the MAVs, retrieve the sample containers, then launch from Mars orbit to return to Earth. The first stages of an MSR are tentatively scheduled for 2014, but many technical and biological decontamination challenges still remain to be met. COURTESY NASA.

The 1971 movie *The Andromeda Strain*, based on the best seller by Michael Crichton, presented a worst-case scenario about bringing an extraterrestrial organism to Earth. In reality, the threat of contamination by alien life is one that NASA takes extremely seriously, not only to protect Earth from potential Martian life-forms, but also to protect Mars from being contaminated by Earth microbes. COURTESY THE KOBAL COLLECTION.

basis, that's all easy compared to ensuring the safe passage of the sample container back to Earth. Statistically, we're required to mathematically demonstrate we have less than a one-in-a-million chance of releasing any returned sample material once we come back to Earth.

"So, we've got this little entry vehicle that has to come screaming into Earth's atmosphere at tens of thousands of miles an hour and land—perhaps even crash-land [as the Genesis capsule did in September 2004]—in the desert and not release any of its contents. We have to demonstrate that there is effectively no chance of that happening. And we've got to demonstrate that everything we've done along the way from when we acquired the samples, put them in the canister, sealed it, and transferred it into the return vehicle—we haven't contaminated the *outside* of that vehicle.

"It's a technological problem that just boggles the mind."

To add a further complication, if one of NASA's Mars missions scheduled to launch before 2014 does happen to uncover unambiguous evidence of current life on Mars, then it could very well be that the first Mars Sample Return mission will be postponed until an absolutely risk-free method of studying that life can be devised. This might be achieved either through complex robotic labs sent to conduct those studies on Mars, or perhaps even by having Martian samples delivered to scientists working at the lunar outpost, which is also part of NASA's Vision for Space Exploration.

From John Wyndham's marauding plants in *Day of the Triffids* to Michael

# NASA'S PRIME DIRECTIVE

*Planetary Protection*

All missions that leave Earth for another body are bound by an international agreement to control the potential to contaminate other worlds (including comets and asteroids) with terrestrial organisms. The process of defining and implementing the requirements of this agreement is called Planetary Protection. These rules also address the return of samples from space to Earth, and hence the contamination of Earth by extraterrestrial organisms.

The rules NASA uses are contained in a NASA Planetary Protection requirements document that is derived from the international COSPAR Planetary Protection Policy. (COSPAR is the Committee on Space Policy and Research.) The preamble of that policy sets the context and importance of this aspect of space missions. In one sense this is a twenty-first-century version of *Star Trek*'s Prime Directive.

**Preamble**

Noting that COSPAR has concerned itself with questions of biological contamination and space-flight since its very inception, and noting that Article IX of the Treaty on Principles Governing the Activities of States in the Exploration and Use of Outer Space, Including the Moon and Other Celestial Bodies (also known as the UN Space Treaty of 1967) states that:

States Parties to the Treaty shall pursue studies of outer space, including the Moon and other celestial bodies, and conduct exploration of them so as to avoid their harmful contamination and also adverse changes in the environment of the Earth resulting from the introduction of extra-terrestrial matter, and where necessary, shall adopt appropriate measures for this purpose (UN 1967).

Therefore, COSPAR maintains and promulgates this planetary protection policy for the reference of spacefaring nations, both as an international standard on procedures to avoid organic-constituent and biological contamination in space exploration, and to provide accepted guidelines in this area to guide compliance with the wording of this UN Space Treaty and other relevant international agreements.

**COSPAR Guidelines for Planetary Protection Categorization**

- IVa: For lander systems not carrying instruments for investigations of extant Martian Life, such as Mars *Pathfinder* and Mars Exploration Rovers.

- IVb: For lander systems designed to investigate extant Martian life. Only the *Viking* landers are examples of this type of mission.

  - Entire landed system must be sterilized at least to *Viking* post-sterilization biological burden levels, or to levels of biological burden driven by the particular life-detection experiments, whichever is more stringent.
  - Subsystems which are involved in the acquisition, delivery, and analysis of samples used for life detection must be sterilized and a method of preventing recontamination is in place.

- IVc: For missions which investigate "special regions" even if they do not include life-detection experiments, all the requirements of IVa apply along with:

  - Definition of "special region": A region where terrestial organisms are likely to propagate OR a region which is interpreted to have a high potential for the existence of extant life forms.
  - CASE 1: If the landing site is within the special region the entire landed system shall be sterilized to at least *Viking* post-sterilization levels.
  - CASE 2: If the special region is accessed through mobility, either the entire landed system shall be sterilized to *Viking* post-sterilization level OR the subsystems which directly contact the special region shall be sterilized to these levels and a method of preventing their recontamination prior to accessing the special region shall be provided.
  - If an off-nominal condition (such as a hard landing) would cause a high probability of inadvertent biological contamination of the special region by the S/C, the entire landed system must be sterilized to the *Viking* post-sterilization levels.

The Mars *Phoenix* lander and Mars Science Laboratory (MSL) rover are currently working to this IVc requirement.

From the planetary protection point of view a number of things are very different and much more challenging for MSL compared to past Mars lander/rover missions. First, there is the identification of a special region concept and the need to deal with "off-nominal" landings. The other is the Mars Global Surveyor and Mars Odyssey observations and measurements. The latest scientific interpretations of those measurements indicate the possibility of water ice being present over a large portion of the Martian surface relatively closer to the surface than previously thought. These factors along with the possibility of MSL carrying a radioisotope power source, which represents a "perennial heat source (the heat source has an 84-year half-life)," could create its own special region where liquid water could provide a habitat for terrestrial microbes to grow. The probabilistic analysis of this scenario is currently an active area of research and negotiation between the MSL project and the NASA Planetary Protection Officer. The results of this assessment will likely have a profound impact on future Mars exploration, in terms of the costs and work needed to establish and implement a particular degree of contamination control for a particular mission and target. Missions will certainly carry more sophisticated instrumentation to more interesting places in the search for water and signs of Martian biology. The measures associated with protecting Mars and scientific equipment from biologic and organic contamination will invariably add to the complexity, mass, and cost of already very complex machines. But the integrity of our current Mars science program demands it as does our responsibility to the future exploration of other worlds.

Crichton's blood-crystallizing supergerm in *The Andromeda Strain*, science fiction is full of object lessons about what might happen in the event of alien contamination. Considering that in the 1960s NASA spent $16 million to construct a Lunar Receiving Lab to safely examine samples from the airless, waterless, radiation-bombarded Moon, which was very unlikely to harbor any form of life, we can be sure that NASA remains just as aware of those lessons today.

But there's another aspect to future missions to Mars that is just as significant as the hardware NASA sends and the tasks that hardware must perform—and that's the question of *where* on Mars do we go next?

It's time to visit again with a familiar scientist who believes the time is right to revisit a place we've already been.


## THE PRESERVATION POTENTIAL

To MER Principal Investigator Steve Squyres, the most exciting aspect of discovering that life once existed on Mars—or continues to exist today—is that in the absence of extensive tectonic activity, the record of how that life formed may still be contained within the rocks of Mars.

"Based on what we know about Mars right now, Meridiani can't be beat. It is *the* place to go."

Meridiani is where the rover *Opportunity* landed, bouncing and rolling to its incredible, interplanetary "hole-in-one" inside a small crater that revealed the sedimentary rocks that pointed to the presence of open water in the Martian past.

And why does that make Meridiani *the* place to go on Mars?

"Preservation potential is the key," Steve says, voice tinged with excitement at what the future may hold. "We have compelling evidence that there was liquid water there. But it's not just that there was an aqueous environment—it's that there is a process going on there that has the capability to preserve, for billions of years if necessary, at a molecular level—*molecular level*—evidence of what was *in* that water."

Halfway around Mars in the Gusev Crater, the *Spirit* rover also found evidence that water had been involved in the formation of local rocks, but in a different way—one that does not necessarily mean that there was open water.

Steve explains why one site offers a more compelling case for a return visit than the other.

"A bunch of sand and basalt boulders at Gusev? What are you going to bring back? You're going to bring back a handful of sand, okay? But the beauty of Meridiani is, not only was there undeniably liquid water there, providing what would have been a habitable environment, there is a mechanism for *preserving* what was in that water.

"If you ask yourself, What are the kinds of deposits on Earth that best preserve evidence of life on Earth, two, three, three and a half billion years ago? It's a 'precipitate.' [That is, rocks created by minerals that were originally dissolved in water.]

Because not only does the water provide the medium in which the stuff can live, but it then precipitates the minerals—very rapidly in some cases—and those minerals can preserve, like bugs in amber, the stuff that was in the water." Then Steve adds a geological detail that suggests an even more exciting possibility to be explored at Meridiani. "In fact, most concretions are actually nucleated around organic stuff."

To shift into plain English for a moment, Steve means that on Earth, minerals similar to the ones at Meridiani got their start when tiny particles in water began to stick together around material of biological origin. As for what a "concretion" is, remember the images of thousands of small spheres scattered all over the Meridiani site—what Steve dubbed Martian "blueberries"? *Is* there any chance at all that the "blueberries" have some sort of organic component in their centers?

Steve's expert, scientific answer is, "Sure! Of course! That's the point of going back!" he adds. (The 2004 MER rovers don't have the instruments necessary to examine the blueberries in enough detail—either chemically or visually—to detect any possible organic material they might have formed around, but the MSL rover probably will.) "Bring them back!" Steve says. "Slice them open! Get them into the best laboratories on Earth!"

Preserved in those Meridiani samples, if there is or was life on Mars, could be microscopic fossils from the earliest stages of life's development—microfossils that no longer exist on Earth today.

The other advantage Steve sees for planning a return visit to Meridiani is that it's proven to be safe. "It's smooth. It's flat. The winds are low. Just from a landing-safety perspective, it's *the* place to go."

On Earth, sedimentary rocks are among the most common places to find fossils. If life ever existed in the Martian lake or ocean in which the sedimentary rocks in this image were formed, it's possible that evidence of that life could be detected by careful chemical and microscopic examination. Though the Mars Exploration Rovers were not equipped with the scientific instruments to carry out a direct search for indicators of past life on Mars, the next landers will be—Mars Phoenix Lander in 2007, and the Mars Science Laboratory in 2009. COURTESY NASA/JPL.

This microscopic image from the *Opportunity* rover reveals the challenges of looking for evidence of past life on Mars. Some observers suggest the vaguely cylindrical, segmented structure near the center of the image resembles an early Earth life-form called a crinoid. Crinoids are marine animals that remain rooted in one spot, feeding on particles of food that drift past their arms. They first appear in Earth's fossil records about 440 million years ago, indicating they took about 3.5 billion years to evolve from the very earliest forms of life. Though early Mars is believed to have been as habitable as Earth, it's unlikely that this condition lasted 3.5 billion years. Thus, if life arose on Mars and evolved at a similar rate as on Earth, there would not have been time for anything as complex as a crinoid to evolve. Another argument against this structure being some form of fossilized Martian life is the fact that there is only one. On Earth, crinoids covered vast fields on the ocean floor, and their fossils are found in great abundance. COURTESY NASA/ JPL/CORNELL/UNITED STATES GEOLOGICAL SURVEY.

Given the challenges of a Mars Sample Return mission, even to a safe landing site like Meridiani where intriguing materials wait to be studied in detail, is there an alternative? With all the scientific and technological advancements of the decades since the Viking missions, and with the Mars Science Lab already being designed to carry state-of-the-art instruments for detecting more varied biologic processes than those on Viking, would the money allocated for MSR be better spent on sending more instruments to Mars, instead?

Matt Golombek addresses that question.

"First of all," he explains, "you *can't* answer that, because you don't know what the technology's going to be able to do for you in ten years. And scientists are the wrong people to ask that question, anyway. Given how difficult and expensive it is to operate the spacecraft on the surface and do even basic rudimentary tasks, some people think that we will never get to a sample-return capability."

But then, as a scientist used to looking at both sides of an issue, Matt counters his own bleak assessment.

"But on Earth, we have incredibly sophisticated instruments that take up a whole room, that we don't know how to miniaturize. And since there's not enough money going into miniaturizing, it doesn't look as if we're ever going to get to the point we can send those instruments to Mars. So getting well-selected samples back on Earth remains a high priority."

What, exactly, is a "well-selected" sample? It's one of those surprisingly simple terms that leads to some fairly complicated new questions.

Steve Squyres would argue for a Mars Sample Return mission to Meridiani in which a rover with sophisticated robotic limbs would be able to scoop up blueberries as well as take core samples of sedimentary rocks. Scooping is definitely within the realm of what is technologically possible today for a Mars rover. (Viking did it and Phoenix plans to.) Coring specific sections of rock, for now, is not proven, though that capability is planned for the MSL mission.

That's why some early versions of an MSR—especially the ones with the lowest price tags—are termed "grab and go." In a grab-and-go mission, the vehicle lands, grabs whatever rocks or other materials that are within its reach, and sends it back.

But Matt doesn't find that scientifically useful. As he bluntly puts it, "We already have 'any' rock—twenty-six of them." He's referring to the twenty-six meteorites known to have come to Earth from Mars—a completely random assortment.

But, in the end, Matt concedes that there's "a disagreement" in the scientific community. On second thought, make that "a fight."

"Some community members feel any sample would do, and others feel that you really need to select samples you can put in context."

On that topic, Steve Squyres and Matt Golombek are in agreement. Whether looking for signs of life or the story of Martian geology, it's not enough to just have samples to study—they have to be specific samples taken from specific places.

"The whole basic science objective is to go read the rocks," Matt sums up.

So how likely is it that a Mars Sample Return mission will launch in 2014?

Firouz Naderi examines the possibility.

"So far, Mars Sample Return has been a movable target, and right now, the politically accepted date is early next decade." To achieve that goal, though, Firouz says there's still a great deal of work to do.

"First, there are technological challenges. If you take a look at Mars Sample Return, several elements of it stand out. First, you have to get the sample off the surface so you do need some kind of a rocket—a Mars Ascent Vehicle, or MAV.

"Then you have the tradeoff of either making the MAV sophisticated so that it has an intelligence to go into a very precise orbit—and that makes it heavier and hard to get off the ground. Or to just take off and put a soccer-ball-sized container, which contains the sample, somewhere in orbit of Mars. That then puts the burden

## IT CAME FROM THE MOON!

*A True Tale of Germs in Space?*

An unexpected outcome of the exploration of Earth's Moon had an important effect on how NASA conducts its exploration of Mars.

On April 20, 1967, the *Surveyor 3* robotic spacecraft landed on the Moon, where it conducted a series of experiments to examine the composition of the lunar soil, as well as taking more than 6,300 photographs.

Two years and seven months later, the *Apollo 12* lunar module landed within sight of *Surveyor 3,* and astronaut Pete Conrad was assigned the task of removing parts of the *Surveyor* lander for study back on Earth, so that the effect of the lunar environment on the spacecraft could be measured.

During the Apollo program, all the materials that were brought from Earth's Moon were processed through the Lunar Receiving Lab (LRL) at Johnson Space Center in Houston. This was a precaution against the remote possibility that there might be strange microbes on the Moon that might affect humans.

The *Apollo 11* samples had yielded no evidence of lunar microbes. But to the surprise of the LRL staff working with the *Apollo 12* samples, one particular culture came up positive.

One possible explanation was that someone assembling the *Surveyor 3* camera—almost three years earlier—had sneezed, and some microbes from that sneeze not only made their way into the camera, they were still there, and still viable! Unless the microbes were the result of contamination within the Lunar Receiving Lab, they had survived more than two and half years of

being in hard vacuum on the Moon, and had been subjected to extreme swings in temperature, from 273°F during the two-week-long lunar day, to -274°F during the equally long lunar night, all the while being bombarded with intense cosmic radiation, unfiltered by atmosphere or the Earth's magnetic field.

No one had ever suspected that Earth microbes could be so hardy surviving under such deadly conditions, and NASA quickly realized that extreme caution would have to be used when sending probes to worlds like Mars, where there was a distinct possibility that Earth microbes might do more than survive. They might spread.

Thus, NASA has always recognized the need to clean any spacecraft intended to land on Mars. The most rigorous example is the first U.S. Mars lander, *Viking.* The *Viking*'s team, driven by their scientific objective to look for extant life, went for full sterilization. Once the spacecraft (both landers, separately) were sealed within their bioshield (to prevent recontamination during handling and launch), they were each subjected to a sterilization cycle, in which their temperatures were raised to 230°F and held there for 30 hours. This killed off most, but not all, of the microbes on board. The final estimate was that there might be about 27 viable spores somewhere, deep within the spacecraft.

Subsequent Mars landers, *Pathfinder,* and MER have used a less strict protocol since they are not looking for life. They were allowed to carry 500,000 spore-like microbes to the surface. For reference, your kitchen table, right after normal cleaning probably has about 100,000 to 10,000,000 microbes on it. But Mars is not a very hospitable place for terrestrial bugs, and unless they can find liquid water, nutrients, and an

energy source and like extreme cold, and can survive extreme UV radiation, they don't stand any chance of reproducing. So when the next generation of more sophisticated biochemistry and maybe even life-detection-experiments next go to Mars, it is NASA's intent, to be sure, that any signs of life we might detect, past or present, will be certain to be indigenous, not microscopic invaders from Earth.

**Astronaut Charles "Pete" Conrad examines the *Surveyor 3* probe on the Moon during the *Apollo 12* mission in 1969. COURTESY NASA.**

on sending a vehicle from Earth to go and search for the sample container, find it and rendezvous with it, then capture it."

Despite the complexity of that scenario, Firouz confirms that it is the one NASA's studies have concentrated on. "A simple MAV to put the sample in orbit, and then a rendezvous-and-capture mechanism to capture this thing and come back. So far these are the two major challenges: the surface activities to obtain the sample, and then the rendezvous and capture, without much human intervention." (In another attempt to reduce the risk for a sample return mission, the Mars Telecommunication Orbiter [MTO], proposed to launch in 2009, is planned to carry a rendezvous experiment to demonstrate two concepts—visual and radio-based—for finding an object in Mars orbit.)

But then, Firouz brings up another critical issue of a potential Mars Sample Return mission. "Planetary protection." And not only for Earth. It's just as important to protect Mars from Earth contamination, to be sure we don't contaminate our samples with bugs we've brought there from our own planet. What Viking did to reduce the likelihood of contaminating samples going into the landers' life-detection instruments was to sterilize (as close as possible) each lander entirely! This was a complex and expensive effort. Indeed, sterilizing Martian landers today could add many tens of millions of dollars to the cost of each mission.

Protecting Earth is also of supreme importance. As Firouz puts it, "After all, if you are looking for biological potential, you have to assume that there is some. So whether you would park the sample somewhere outside the atmosphere of the Earth and then go fetch it, or have it make a direct entry, you have to make sure that you talk about the containment and then the initial quarantine of the sample."

Another critical issue is, of course, financial. "This is a two- or three-billion-dollar mission." That price tag is so high that one strategy calls for breaking up Mars Sample Return into several Mars Exploration Rover–class missions. That way, other rovers—perhaps even the MSL— could land on Mars, explore different landing sites, do experiments, and as one of their objectives, gather samples that would be picked up by another mission, twenty-six months and several fiscal years later.

"But I must say," Firouz concludes, "that the science community's emphasis on the sample return has been constant. Every time we have asked it—and we have asked it a little bit different each time to make sure that they really meant what they said the last time—the answer always comes back the same: Yes."

So, while the 2014 launch date might not yet be written in Martian rocks, sometime in the not-too-distant future, for the first time a mission won't involve just something from Earth going to Mars, but something from Mars coming to Earth.

And once scientists have those Martian samples in hand and critical questions can be answered about the possible existence of life and the potential toxicity of exotic Martian dust and chemical compounds, it will be time for one more small step, one more giant leap.

It will be time at last for humans to make the journey to another world, and go to Mars in person.

# CHAPTER EIGHT

# "It Can't Happen Soon Enough for Me"

## Humans on Mars

*"Ask ten different scientists about the environment, population control, genetics—and you'll get ten different answers. But there's one thing every scientist on the planet agrees on: whether it happens in a hundred years, or a thousand years, or a million years, eventually our sun will grow cold, and go out. When that happens, it won't just take us, it'll take Marilyn Monroe, and Lao-tzu, Einstein, Maruputo, Buddy Holly, Aristophanes—all of this. All of this was for nothing, unless we go to the stars."*

—J. MICHAEL STRACZYNSKI, *BABYLON 5*

## BUILDING THE FOUNDATION

In 1950, eight years before NASA was formed, the first spacecraft intended to take humans to the Moon blasted off with a crew of four men and one woman.

Unfortunately, the redoubtable crew of Rocketship X-M ran into some engine trouble, missed the Moon and, as depicted in the movie of the same name, landed on Mars instead.

If only it were that easy.

Today, of course, filmmakers are well aware that a flight to Mars takes six months or more. But in their movies about going to Mars, they still take the easy way out by simply cutting from the excitement of the launch to the excitement of the landing, skipping those six months of cruise when presumably nothing happens. (Though in the 1950s, Mars-bound crews could count on at least one meteor shower to bring some excitement to their day.)

But in real life, no cutting is allowed, and using propulsion technologies that are available today, the first human expedition to Mars could very well require astro-

As odd as this looks to us today, in 1950, when the movie *Rocketship X-M* was released, it was considered within the realm of possibility that all a human would need to survive on Mars was warm clothes and an oxygen breathing mask. COURTESY THE KOBAL COLLECTION.

nauts to spend three years away from Earth, and less than half that time would be spent on Mars.

When this mission will take place, no one at NASA knows for sure. The uncertainty goes beyond any questions about how to find the money, or how to develop new technology. Simply put, right now, late in the year 2004, no one knows exactly how to do it.

Mars activists will claim that's nonsense. NASA has demonstrated the capability to reliably send spacecraft to Mars, to have them orbit, and have them land. The Russian space program has demonstrated that humans can survive in weightless conditions for longer than the time a flight to Mars will take. The effects of radiation exposure are overblown. The activists claim that the only thing preventing humans from getting to Mars in the next ten years is the lack of political will to spend the billions of dollars necessary to accomplish the task.

And this is where the activists' position unexpectedly agrees with that of NASA and its 2004 Vision for Space Exploration.

Think about that time frame for a moment. The Mars activists maintain that, even with full funding, it will take ten years to ready a human expedition to Mars. That's more than thee times as long as it took JPL to develop, design, and build the Mars Exploration Rovers.

With the exception of the part about a giant face built by aliens, the 2000 film *Mission to Mars* featured the most accurate depiction to date of Mars and the technology that might be used to take us there. COURTESY THE KOBAL COLLECTION.

The difference is, JPL knew *how* to build rovers and send them to Mars, and they knew what conditions would be waiting for the rovers when they arrived. They'd already done it once for Mars *Pathfinder*. The MER mission was the next stage of advancement and improvement.

But humans need far more supporting technology than solar-powered rovers—especially considering those rovers are on a one-way trip. And for all the discoveries NASA has made about Mars since that first flyby of *Mariner 4* in 1965, we still don't know if merely inhaling the dust of Mars—something astronauts will not be able to avoid once they start exploring the surface—will prove debilitating, or even fatal.

Quite possibly, if there were a compelling reason, NASA could mount a human expedition to Mars within ten years. Necessarily, it would be a crash program, racing the clock. And, as MER showed, the only way to beat that clock is to spend money—enough to pay for an enormous workforce and multiple paths of development.

Lacking that compelling reason, the ladder leading to Mars will be climbed one rung at time, with NASA spending taxpayer dollars slowly and carefully, which means thorough planning and execution to ensure there is no wasteful duplication

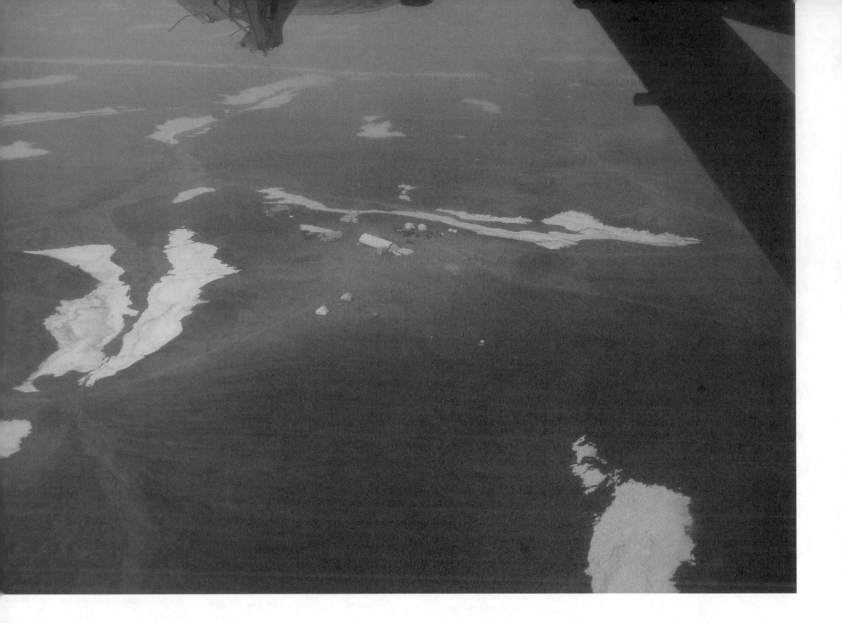

of effort. Yet make no mistake. We're still going to Mars: NASA has been specifically directed to make this its goal.

As of now, late in the year 2004, NASA believes it is reasonable to expect that the proper scientific and technological foundation for planning to send humans to Mars will be achieved around the year 2020.

Here's what that foundation will likely include.

In terms of technology, no significant breakthroughs are required. NASA *does* know how to get spacecraft to Mars and how to build reliable spacecraft systems that will function for years. For safety's sake, a Mars expedition will possibly include several spacecraft, including at least one habitat module that will land automatically on Mars before the crew ever leaves Earth. A Mars-orbiting supply module with the fuel and consumables required for the trip back home might also be put in place before the human mission begins. The often discussed idea of in situ propellant production—producing hydrogen and/or oxygen from Martian water or rocks for use as fuel and oxidizer for the return flight—is also a possibility. But this approach would require finding the right raw material, demonstrating a reliable production and storage capability, and then actually producing and storing a vast quantity of material, all by robotic machinery, and well before humans leave for Mars.

In terms of a human's ability to remain healthy over the course of a three-year mission, most of the necessary medical research has already been accumulated by the Russian space program, and by studies done on the International Space Station. Radiation shielding for the crew's spacecraft and Mars surface habitats will be crucial, for solar flares can be fatal outside of the Earth's protective magnetic field. Seven months in weightless conditions on the cruise to Mars will likely degrade the crew's performance upon arrival, but not critically so. Landing will probably be accomplished by the same automated system used to set down the advance habitat modules.

In terms of Mars itself, extensive mapping of the planet must be completed in order to choose the best site for humans to land. Like all landing sites, it must meet stringent requirements of safety and offer significant opportunities for scientific investigations. Without question, the chosen site will have been explored in advance by rovers. And, by the time of the first expedition, so the astronauts on the surface can be in constant touch with Earth and provide a constant flow of information, orbiters must also be in place to serve as a permanently accessible Martian communications network.

Habitat modules . . . rovers . . . orbiters . . . see how the foundation missions begin to add up? (Though with the Mars Reconaissance Orbiter, the Mars Science Laboratory, and the Mars Telecommunication Orbiter missions, we are already starting to build and demonstrate how that infrastructure can be put into place when the time comes.)

But mapping and rover exploration will not be enough to ensure the safety of the first human explorers. Samples of Martian dust and rocks have to be carefully analyzed—most likely back on Earth—to determine if they pose a threat to human health. Today, the fine powder observed on Mars has raised the possibility that inhaling it could cause a form of pneumoconiosis, also known as "miner's lung."

And there is another "wild card" possibility to consider, one that could postpone human exploration by decades. What if Martian life *is* discovered by robotic missions? If so, then it's not likely that no humans will set foot on Mars before intensive robotic investigation of that life can assess its potential danger to human health and life in particular, and to Earth life in general.

Consider, too, that if life is found, even the most primitive microbial forms are likely to exist as thousands of different species. Some Earth microbes have no effect on us, and others cause terrible disease. For the safety of the astronauts, we would have to know if the same is true of Martian life. To thoroughly understand the nature of that life, dozens if not hundreds of different sites would have to be studied, with samples taken from each of them. All of which means more foundation missions, more money, and a great deal more time.

But all these potential hurdles will not divert NASA from its goal.

Because, in the end, going to Mars is a journey that's important for everyone on Earth.

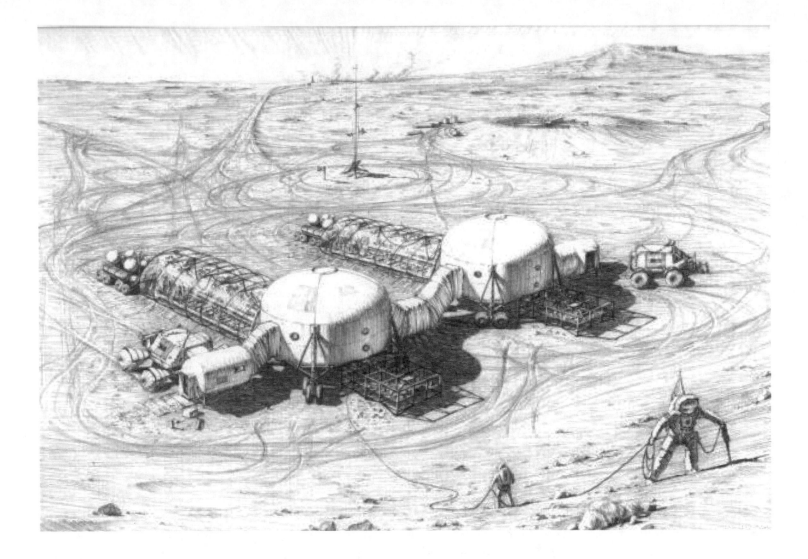

NASA's first human base on Mars will be prepared years ahead of the crew's arrival, to ensure that ample supplies of food, water, and fuel for the return trip are already in place. COURTESY NASA/JOHNSON SPACE CENTER.

## DOING THE RIGHT THING

Sending humans to Mars is a practical way to continue our exploration of the solar system. Steve Squyres, principal investigator for the Mars Exploration Rover (MER) mission, who has witnessed firsthand the success of robotic investigation of Mars, has this to say about plans for "the human initiative."

"It can't happen soon enough for me." He quickly adds, "There's nobody alive who's a bigger fan of sending robots to explore Mars than I am. But I believe that the most comprehensive, the most successful, and frankly the most inspiring exploration of Mars will be done by humans."

And on this subject, Steve is speaking from experience.

He describes what he calls "an interesting experiment" he once conducted. "A few years ago, we were trying to figure out how to use what became the MER rovers to do fieldwork on Mars. We had this concept for a rover: we've got the mobility, we had the map, we had cameras, we had the arm, we had the instruments. But how do you actually use the thing to do field science?

## STAR TREK ON MARS

*Choosing a Crew Worthy of the
Starship Enterprise*

When the first human expedition to Mars lifts off from Earth, how many people will be on board, and what kind of people will they be?

Astronautical engineer Robert Zubrin, a leading advocate for traveling to Mars as quickly as possible, has given the matter careful thought. In fact, he has developed a complete plan for safely and inexpensively beginning the human exploration of Mars, which he calls Mars Direct.

Mars Direct crews will spend two and half years away from Earth: six months of travel time going to Mars, a year and a half spent exploring the Red Planet, then another six months of travel time returning to Earth. Factoring in requirements for food, water, oxygen, and fuel for transportation, Zubrin has determined that the optimal number of crew members for a Mars Direct mission is four.

And he has a colorful way of describing just who he thinks should be in that crew.

Zubrin has concluded that the most likely reason for failure of a mission to Mars would be because of the unexpected failure of one of its critical systems, such as propulsion,

flight control, or life support. Thus, he believes the most important member of the crew will be someone with the skills of a mechanic.

Then, since the purpose of going to Mars is to gather scientific data, if that data can't be gathered, then the mission is also a failure. Thus, the second most important job is that of field scientist.

In fact, Zubrin believes that the mechanic and the field scientist are so important to the success of the mission that the ideal crew will have two of each. As far as piloting and medical skills are concerned, he feels it is easier to train a geologist to fly a computer-controlled landing craft than it is to train a pilot to be a geologist.

Zubrin sums up his crew configuration in *Star Trek* terms. The best crew will have two Scottys and two Spocks, but has no need for any Kirks, Sulus, or McCoys.

Of course, if anyone discovers Klingons on Mars, that crew configuration could be subject to change.

**The ultimate Mars crew?** COURTESY PARAMOUNT PICTURES.
STAR TREK™, ® & © PARAMOUNT PICTURES. ALL RIGHTS RESERVED.

"So we had this rover called FIDO, and four or five years ago we took it out into the field at a place called Silver Lake. And that very first time, the rover broke down. So, I'm sitting out there with a bunch of high-priced geologists and nothing to do. I thought, All right, let's all go for a geology walk. Everybody gets their boots and their rock hammers and their hand lenses and their notebooks, but I just took a notebook and a stopwatch. I had my rock hammer, but I wasn't going to use it.

"So we go out and we walk about a half mile in the desert, and we get to a sort of an interesting ridge where there are a bunch of rocks exposed and everybody else wants to work—you know, 'geologizing,' just looking around. So, when one of them looked off in the distance and saw an interesting rock, he'd walk over to it and he'd reach out to it and he'd hit it with a hammer to expose the interior, then look at it with the hand lens. They were doing all the things that *we* are doing with the MER

## THE MARS-SPANGLED BANNER

*The Mars Flag*

The Mars flag was designed in 1984 by Thomas O. Paine, former NASA administrator and chairman of the National Commission on Space.

Paine wanted to create a symbol that would serve as a rallying point for building support for the human exploration of Mars. The first use he intended for the flag was to make it an award for those individuals who make notable contributions to making Mars a part of our future.

Fittingly, The Planetary Society now presents the Mars flag each year to the winner of the Thomas O. Paine Memorial Award to continue to honor those achievements.

The Mars flag shows a representation of a sliver of Earth to one side, as a reminder of where we came from, and a star to the other side, to remind us of where we are going. In the center of the field is a representation of Mars itself, with an arrow pointing out to the star, acknowledging that Mars is not our destination, merely a way station on a journey that has no ending.

**The Mars Flag, designed by Thomas Paine, being raised on Mars.** ARTWORK BY CARTER EMMART/COURTESY OF THE ARTIST AND THE PLANETARY SOCIETY.

rovers, now: driving, removing outer layers of material, microscopic imaging—all that kind of stuff.

"And I timed them with my stopwatch. How long it took them to walk that far, and break that rock, and to look at that thing."

Steve smiles as he reveals the results of his surreptitious study. "What our rovers, today, do on average in a single Martian sol, those geologists did in about *thirty seconds.*

"Even with the best robots that JPL can build and that money can buy today, we're just a damn long way away from doing what a person can do. We really are."

Practicality is not the only reason Steve supports the idea of sending humans to Mars. "I think there's all kind of reasons to go there with humans. Getting the job done right is one of them. But another one is just that it's going to captivate and inspire people in a significant way that robots aren't going to.

"I think we're doing the right thing by sending the robots in first to answer the basic questions. So that when the humans follow, they'll be able to use their capabilities most effectively."

Again, Steve speaks from his experience with robot-assisted exploration.

"I had an experience with that down in Antarctica, where we were doing some work in the Dry Valleys. This is years ago, the mid-eighties, but I was already thinking that someday I'd like to go to a dry lakebed on Mars and look at sediments there. So, the question is, How does sedimentation happen in ice-covered lakes on Earth?"

To answer that question, Steven took part in a combined, human-robotic exploration of a frozen lake.

"We spent eight weeks in the Dry Valleys, and I got in a grand total of six dives, with a total bottom time of about four hours, in sixty to eighty feet of water. So in eight weeks down there, I spent only four hours actually in the environment that I wanted to work in.

"But we had this little robot, this little remotely-operated vehicle that's really easy to operate. It's got lights, and it's got little thrusters, and it's got video cameras. You just get out on the ice, you set up your tent, you get your cup of coffee, you've got your little console, you've got a diesel generator going, and you could operate this thing for hours!

"What I was able to do, using this low-risk, reliable robot technology, was to become familiar with the environment that I was going to go into. I answered the simple questions, saw what the terrain looked like, knew what I wanted to do. So then, when I suited up and actually went into the environment myself, instead of going down through sixteen feet of ice and flailing around and wasting time, I knew just the spot to go to. I've got a checklist. I've already built my equipment to match the conditions I've seen. I know what kind of grid I want to lay out. Bang bang bang . . . out come the sample cores. I've got my science!

"In only four hours bottom time, I was able to maximize my science return because I'd viewed the environment first with the robot."

Given his passion for the human exploration of Mars, it's no surprise that Steve has an enthusiastic reply when asked what an appropriate timetable for the first human expedition might be.

With a smile, he says, "Hey, send them to Meridiani Planum right now!"

## AN EXPLOSION OF KNOWLEDGE

As a scientist, Matt Golombek lives to pursue knowledge, and though he knows there are two sides to the question of whether the human initiative to Mars should be pursued, there's no mistaking where his passion lies.

"Going to Mars is about exploring and knowing the place you live in. I would

Earth as seen from the Moon. COURTESY NASA.

say one of the crowning achievements of our civilization is that we have given to humanity the knowledge that the Earth is not a single place in the universe. We know what the Earth is and we know how it compares and exists within our solar system.

"It's totally arguable on a separate issue about whether we should be sending people into outer space, whether that's a smart thing to do, or stupid, or if it's worth the money—that's a whole different topic.

"But in our lifetime, Saturn, Jupiter, Venus—all the planets—have gone from being points of light in a telescope to worlds with characteristics we can understand. What an explosion of knowledge!

"I mean that, that's what space exploration's about. It's about understanding what you are, and where you are."

## HOW RARE COULD THIS BE?

Firouz Naderi, director of JPL's Mars Exploration Program, understands the pull of Mars, and the reason why polls show solid support for NASA and the space program, tempered, of course, by legitimate concerns over cost.

For Firouz, that support comes down to one question—*the* question.

"Are we alone?" he says. "That still is the thing that fascinates people. Because you do look at the sky and I think everybody has wondered if somebody is looking back down."

As to public impatience about the time it could take before humans might make the journey to Mars, Firouz feels people aren't aware of the challenges of that goal. And he thinks he knows why.

"The good and the bad about Hollywood is that with the movies, it gets people excited about space. But the movies make space travel so trivially easy."

Firouz says he faces another question in almost every social gathering when his involvement in Mars exploration comes up. "People continually ask, 'Well, how long are the people going to be up there?' Because some people take it for granted that the missions there now have astronauts. People tell me, 'Well, I saw such and such a movie and we went not only to Mars, but to this other planet, somewhere else outside the solar system.'

"So I'm not sure that the majority of the people have a sense of how difficult really it is. Because people get jaded watching the Hollywood movies."

Firouz, himself, is very much aware of the challenges that will have to be met before humans make the journey to Mars. "They asked Wernher von Braun in 1950 when he thought we would go to Mars. And he said 'about a century,' which would make it about 2050. I think he may still be proven right."

In addition to the issues of technology, mission duration, and human health, Firouz also sees another challenge to be faced in going to Mars.

"I rather suspect there might also be a bit of psychological adjustment. You know, when we went to the Moon, you could look back and Earth was there—home,

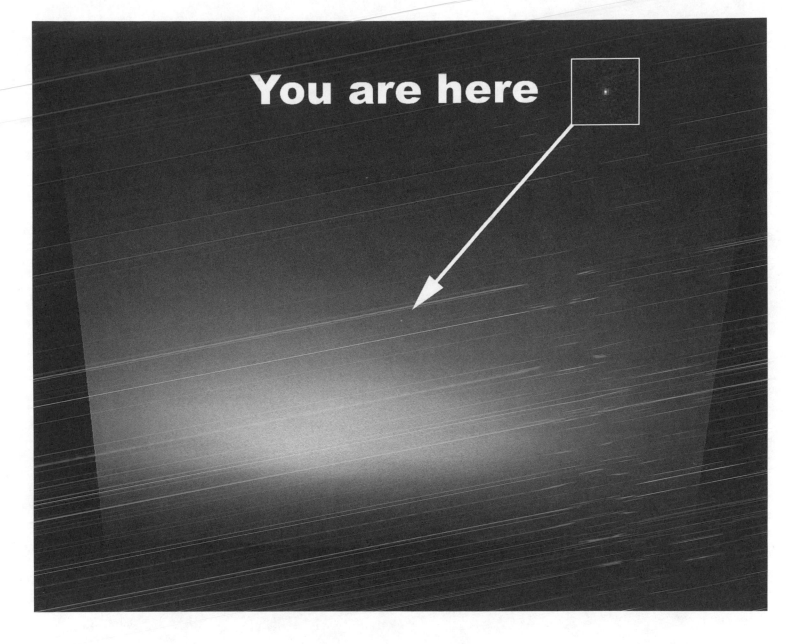

**You are here**

This historic image is the first taken of the Earth from the surface of another planet. It was taken in March 2004 by the Mars Exploration Rover *Spirit* one hour before sunrise on the sixty-third Martian day of its mission. The contrast of the image has been enhanced to make Earth easier to see, otherwise our planet was too faint to be detected. COURTESY NASA/JPL/CORNELL/TEXAS A&M.

big as life, a couple of days away. Now, when we go to Mars, we will see the Earth as a speck in the skies. So there is definitely a sense of separation. I'm sure we will find hardy souls who are willing, maybe even chomping at the bit, to take on that challenge. But in general, there probably is also some psychological effect from that kind of isolation."

Astronaut Sally Ride agrees with Firouz. In an interview she did with Brian Muirhead's daughter, Alicia, Sally said that the greatest uncertainty for a mission to Mars may lie in the psychological preparation and management of the crew. She pointed out that many of the Apollo astronauts who traveled to the Moon came back changed in some way by seeing their home at such a great distance. Though we can test a potential Mars crew in realistically simulated conditions for everything from

## THE DRAKE EQUATION

*Estimating Aliens*

Most people at some time in their lives look up at the stars and wonder if there might be intelligent life on other worlds looking back. But some scientists aren't content to merely wonder about such things. They consider SETI—the Search for Extraterrestrial Intelligence—to be a legitimate area of scientific study. And that means they have developed methods of defining the question in order to approach it in a logical manner.

One of the most significant tools used to guide people involved in SETI is the Drake Equation, which was formulated by radio astronomer Dr. Frank Drake in 1961 as a way to estimate the number of advanced civilizations that we might be able to communicate with in our galaxy.

In the Drake equation, $N$ is defined as the number of civilizations in the galaxy with which we might be able to communicate at any given time. The fact that human civilization on Earth has the capacity to transmit radio signals across interstellar distances means that $N$ has to at least equal 1. The big question in SETI is, under what conditions might it equal a number larger than 1? In other words, does intelligent life exist somewhere other than Earth?

Here's how the Drake equation can be used.

$R$ = the average rate at which stars are born in our galaxy. Based on recent findings of the Hubble Space Telescope, which has photographed "star cradles" where stars are being born, most astronomers accept an approximate value of 20 stars per year for this figure.

$fs$ = the fraction of stars that are the right size and temperature to be life-giving suns like our own. Based on extensive spectroscopic surveys of the galaxy's stars, most astronomers estimate this fraction to be 10 percent, or 0.1

$fp$ = the fraction of those suitable suns that have planets. The recent quickening pace of discovering new planets around others suns makes some astronomers think this fraction could be as high as 50 percent, or 0.5.

Now the equation goes into areas for which useful astronomical data do not yet exist. That means the values of the terms have to be determined by guesswork.

$ne$ = the number of planets in those solar systems that are located at just the right distance from the sun to support life—that is, where the average temperature allows water to exist as a liquid, which is absolutely essential for life on Earth. In our solar system, some astronomers would argue that $ne$ = 3: Earth, Mars, and the Jovian moon Europa.

$fl$ = the fraction of those planets at a suitable distance from their sun on which life actually develops. If we accept 3 as the value of $ne$, then we know $fl$ equals at least 0.33–of the three suitable worlds in our solar system, one of them definitely has life. If new missions to Mars discover that life either exists, or once did exist on that planet as well, then $fl$ will double in value to 0.66. And if life is discovered in the planetary ocean of Europa, then the value of $fl$ will increase again to 1.

$fi$ = the fraction of those planets with life on which a species evolves with the capacity for intelligence. Some scientists give this fraction an incredibly low value, arguing that on Earth, despite billions of years of evolution which gave rise to millions of different species, only one of those species developed technological intelligence—humans. That could mean

launch to landing, we can't re-create the conditions of separation from Mother Earth, her image, her gravity. How do we test and prepare a crew for being in deep space with no chance of return to Earth for at least three years? We don't really know.

Firouz is also well aware of the most practical challenge affecting plans for a human expedition.

"The last item is the financial. At one time, they pegged the cost of a human mission at four hundred billion dollars which I think is probably an upper amount. And I have heard numbers as low as fifty billion dollars. The range is so wide, but whatever it is—it's a *lot* of money. So it would require a national will to go. Maybe even an international will to go. There needs to be a lot of support for taking on something like this."

that intelligence is extremely rare, and $fi$ could be equal to 1 chance in 10 million or 0.0000001. Other investigators, however, believe that since intelligence carries a high survival benefit, it will always arise wherever life can be found, even if it does take time. If that's true, then $fi$ could be equal to 1. Clearly, with such a large spread in possible values, the Drake Equation can produce many different estimates.

$fc$ = the fraction of those intelligent species who go on to develop the technology and the desire to communicate with other species across interstellar space. Again, this value is open to debate. Humans have the technology to communicate with other worlds, but as a species, we have not yet demonstrated the commitment to do so. No government is willing to spend the money required to build an interstellar transmitter. Some people even argue against announcing our presence to the rest of the galaxy, citing the experience of every less technologically advanced culture on Earth that has come into contact with a more technologically advanced culture. The less advanced culture is invariably devastated and usually wiped out.

And finally . . .

$L$ = the lifetime of an intelligent species with the technology and desire to communicate. The advances in science, which have given us the ability to send signals to the stars, have also given us the capability of destroying ourselves with nuclear and biological weapons. It has been just over one hundred years that our species has possessed the technology to communicate across interstellar space. How much longer will our civilization last? Another hundred years? A thousand? All such assumptions, whether based on actual data or informed guesswork, will affect the values assigned within the Drake Equation, which will, in turn, affect its final outcome.

Here're some examples.

We'll begin by assuming that the first three terms of the equation agree with current astronomical observations. If we further assume that every suitable solar system will have:

one suitable planet ($ne = 1$);

that half those planets will develop life ($fl = 0.5$);

that all of the life-bearing planets will give rise to an intelligent species ($fi = 1$);

of which half those species will choose to communicate ($fc = 0.5$);

and those species' civilization will survive at least sixty years ($L = 60$);

then $N = 15$, which means there could be fourteen other civilizations on other worlds in our galaxy right now, capable of communicating with us.

However, if life is rarer than we think, perhaps arising on only one of a thousand suitable worlds, then even if all the other assumptions remain the same, $N = 0.03$. Under that condition, the Drake Equation suggests we might be the only technologically advanced civilization not only in our galaxy, but in 32 others, as well.

The ongoing exploration of Mars will allow SETI investigators to further refine the terms of the Drake Equation. But if life on other worlds is truly like life on Earth, then the irony might be that our galaxy is teeming with thousands of technologically capable species, all of them, like us, unable to get the funding to make the first call.

(Try the calculation at seti.org/seti/seti_science.)

Is there some discovery we could make that might encourage that support?

Firouz thinks that's a definite possibility. "I believe people don't get up every morning, look at the sky, and say, Are we alone? But somewhere, lingering in our consciousness, is this question of: Is everything else that we see just a backdrop for our existence? It's the ultimate vanity to be thinking that.

"If you take a look at the scientific evolutionary path that takes the universe from a chemical universe to a biological universe, we don't know whether that's such a remote and improbable process and an event that only took place here. Or, given the enormous numbers of stars—if you talk about what Sagan used to say: 'billions and billions'—that even if life is low probability, by the time you multiply by such a huge number, this is not such a rare occurrence.

"I think the sociological, the religious implications, the spiritual aspect of

## TOTAL RECALL

*The Downside of Being Inaccurate*

Science fiction inspires scientists, while scientists provide fresh material for science-fiction writers. Part of that arrangement can be seen in the slow but generally steady increase in scientific accuracy when it comes to portrayals of Mars.

In 1950, two films depicted conditions on Mars in a way that might have seemed plausible at the time. *Flight to Mars* had the first human expedition trudging around the bleak, snow-dusted landscape in heavy flight suits and leather helmets to protect against the cold. *Rocketship X-M* took conditions one step further by making sure the crew wore cold-weather gear *and* high-altitude breathing masks.

Then in 1955 *Conquest of Space* had the first astronauts on Mars properly outfitted in spacesuits and helmets that, to the film's credit, were different in design from the spacesuits worn in the same movie by astronauts working in space. Unfortunately, one of the scientists checks out the Martian soil by taking off his glove and scooping it up with his bare hand. Makes you wonder what the filmmakers thought the helmets were for.

Accuracy was not a significant factor in most Mars films to follow. In 1965 *Robinson Crusoe on Mars* was a notable exception, with at least an honest attempt made to explain how a marooned astronaut could live off the land on Mars—as it might have been prior to the discoveries of the *Mariner 4* flyby, that is.

In terms of capturing the look of the Red Planet one of the best films is *Mission to Mars,* made in 2000, featuring technology based on actual NASA concepts and wonderful attention to detail in the creation of Martian landscapes. The level of scientific accuracy and logic in the script, though, did not measure up to the art direction.

Ten years before *Mission to Mars,* another Mars film had the odd distinction of being the first movie to employ one of the biggest scientific ideas about Mars and then present that idea in the most inaccurate way imaginable. The film is the Arnold Schwarzenegger blockbuster *Total Recall,* and the big science idea is "terraforming."

Terraforming refers to the audacious idea of transforming another planet to make it suitable for Earth life. The concept originated in the 1930 science-fiction novel *Last and First Men,* by Olaf Stapledon. The term itself was coined by science-fiction writer Jack Williamson in 1942. Today it is a legitimate subject of interest to thousands of scientists who study climate, geology, biology, and astronomy, and the topic is most often associated with the planet considered the most likely prospect for terraforming—Mars.

In basic terms, the idea is simple. Billions of years ago, Earth couldn't support the type of life it has today. There was no oxygen in the atmosphere, no ozone to protect the surface from lethal levels of ultraviolet radiation. A time traveler from today who visited the Earth three billion years ago would die as quickly as an astronaut without a spacesuit on Mars today.

So, obviously, Earth terraformed itself. Microscopic plants evolved that used carbon dioxide and produced oxygen as a waste product, and then other life-forms evolved to utilize the oxygen and so on, and so on. The atmosphere became thicker, and the ozone layer formed, making it possible for life to leave the protection of the ocean.

Meanwhile, just another few tens of

whether life is possible anywhere else, that could be the imperative that builds public support. If we find evidence of life on Mars, then forget about the billions and billions. If at the first place we look, the next rock away from us from the Sun, we find life, then how rare could this be?

"At NASA, even we in this business, we don't get up every day and ask ourselves the question, Are we alone? But there are moments it just sort of hits you. There are occasions where you just realize what business you're in, and what implications that has, and you get caught up in things that elevate you a little above the everyday routine.

"So I believe there may be triggers to building support, because I believe simmering underneath our consciousness is this desire to know what it all means, what's the context for our being here. I think a little spark of discovery could help build support. But given all the challenges, it's going to take something monumental." To confirm Firouz's perception, a significant part of the interest and ex-

millions of miles farther out from the Sun, Mars became cold, bleak, without much of an atmosphere, completely unsuitable for all but the simplest and hardiest of Earth life-forms.

Which raises the question: Could we do deliberately to Mars what happened naturally on Earth?

The trick to making Mars suitable for Earth life is to make the atmosphere thicker, with about the same percentage of oxygen as on Earth—20.9 percent. Scientists have come up with two different ways it might be possible to achieve those results, though both require technology considerably more advanced than any available to us today.

One scheme involves constructing vast mirrors in orbit around Mars, hundreds of miles wide, to concentrate reflected sunlight on the Martian ice caps. Over centuries, the frozen carbon dioxide, then frozen water will evaporate, helping to thicken the atmosphere and warm the planet. A slightly faster plan involves strapping rockets of unimaginable power to 10 billion-ton asteroids and crashing them directly into the Martian ice caps. Then, whichever scheme is used, simple

plant life from Earth would be seeded across the planet to begin to manufacture oxygen.

There's no doubt the process would take millennia. But scientists study the concept for what they can learn about how the Earth developed the characteristics it has today. Our ongoing study of Mars might also reveal why what was once, apparently, a warm wet world became a cold dry one. In other words, by studying the concept of terraforming, we might learn if what happened to Mars could happen to us.

That's the other important fact to keep in mind before dismissing the idea of terraforming out of hand. In time, Earth will die. Hundreds of millions of years from now, our Sun will become larger and hotter, and eventually Earth will be unable to support life. However, as the Sun's heat increases, Mars will once again become a warm world, and in that far future time, if there is still a form of intelligence in our solar system, the Red Planet could very well be the new home of whatever we humans have become millions of years from now, or of whatever species has replaced us.

It's a fascinating, huge idea, and one that's been part of many great science-fiction

novels, most particularly the award-winning trilogy by Kim Stanley Robinson—*Red Mars, Blue Mars, Green Mars*—which wonderfully chronicles the multigenerational saga of the terraforming of that world.

Which brings us back to *Total Recall*.

In that movie, Mars *is* terraformed, but given that few Hollywood producers are interested in a story that unfolds over centuries, the entire terraforming process—conveniently carried out by vast underground alien machinery—takes place in about three minutes, just in time to save its hero from suffocating in the almost nonexistent Martian atmosphere.

There is some consolation offered at the movie's end, though, when it's suggested that the entire adventure might be nothing more than a dream.

At least in that sense, the movie is completely accurate. For now, living on an Earth-like Mars *is* just a dream.

But just like the dream of simply going to Mars, it too might come true someday.

citement about the Mars Pathfinder and MER missions has come from the initial reports of discoveries of microbial fossils in the Allen Hills meteorite (ALH84001). While still openly and sometimes loudly debated, the possibility of hard evidence of microbial life on Mars clearly has the potential for shaking things up here on Earth.

## FULL CIRCLE

"Remember: Four thousand people did this."

Those were the last words MER Project Manager Pete Theisinger said to us at the end of one meeting, because he didn't want us to forget.

The Mars Exploration Rover missions are a triumph of science and engineering. They drew from the best of Mars *Pathfinder* and NASA's other great achieve-

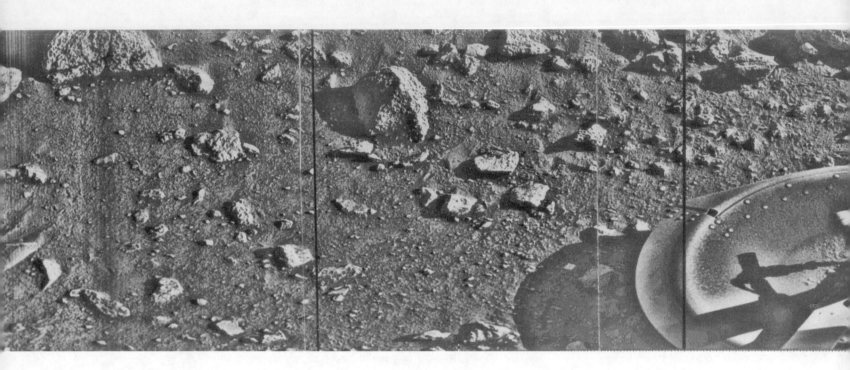

This historic photograph taken by the *Viking 1* lander is the first image sent back to Earth from the surface of Mars. Thousands have followed, with more to come. COURTESY NASA/JPL.

ments, to set a new standard for all the Mars missions to come—missions that one day, inevitably, will include humans.

But even as we marvel at the technology and applaud those five or six people who appear on the podium for press conferences, let's not forget what Pete said.

"*Four thousand* people did this."

Brian Muirhead understands those words. On the morning of July 6, 1997, the day after the *Sojourner* rover rolled out onto Mars, he stepped onstage at the Planetary Society's Planetfest gathering of space enthusiasts in Pasadena, California, and was overwhelmed by the applause and the standing ovation he received as the representative of all who had worked on the mission.

That experience and what it taught him remain with him today. "People out there really care about what we did here. These missions are not just a JPL or NASA or American achievement. They're a human achievement."

In that audience that day, Judith and Garfield Reeves-Stevens were on their feet applauding, participating in the adventure.

Though we were yet to meet as co-authors, the message that day was clear to the three of us—the same message that fills the pages of this book.

Whether scientist, engineer, or one who supports them, each of us with a different story, yet all on the same journey—together, we *are* going to Mars.

In our hearts, in our dreams, we're already there.

# A F T E R W O R D

IN DECEMBER 1972, during the *Apollo 17* mission, I had the opportunity to do what no other geologist had ever done: fieldwork on another world, the surface of the Moon. When I left the Moon with fellow astronaut Gene Cernan, the first phase of humanity's personal exploration of other worlds came to an end. In the more than thirty years since, no human has ventured farther from Earth than the few hundred miles it takes to achieve the low orbits of the Space Shuttle and International Space Station.

It's time for that to change.

As the world's leader in space exploration, NASA finally recognizes this. The President's and the Agency's Vision for Space Exploration sets the goal for this nation to once again explore space and extend a human presence beyond Earth and into our solar system. I like to think of it in even larger terms—not just a human *presence,* but actually moving civilization into the solar system: to the Moon, to Mars, and, as the president says, beyond.

Left unstated in NASA's new vision is specifically how the roles of humans and robots will be balanced. As a human explorer, I am excited by this challenge. But, at the same time, as an Apollo astronaut and a geologist, I am disappointed when the media report that a few, generally nameless "space scientists," are opposed to human exploration of the Moon and Mars. Certainly, that is not the spirit of exploration expressed in these pages by the engineers and scientists who are at the forefront of the robotic exploration of space, looking forward to the day when humans will follow.

It's my belief that most critics of the human exploration of space are strangely unaware of the extraordinary foundation of scientific knowledge about the origin and history of the Moon, Earth, planets, and solar system that resulted from exploring the Moon in person. This immense body of data and understanding could not be obtained by robotic exploration at anywhere near the same speed or even at a comparable cost; quite probably, much of it could not have been obtained at all. It provides the foundation for the interpretation of scientific findings from most of the robotic missions of exploration that followed Apollo, and those still to come. Relative to a specific mission, sending human explorers back to the Moon and to Mars and beyond may indeed cost billions of dollars more than pure robotic missions, but to say that it would add "not much in the way of discovery" shows ignorance of the unique scientific legacy of human exploration of the Moon.

For deep space, exploration should always use the best combination of human and robotic missions, tools, and techniques. Robots clearly have enormous value for ventures on the Moon, Mars, and the asteroids. Any task that can be automated at a reasonable cost should be left to robots, particularly with respect to repetitive data collection, deployment of scientific instruments intended to remain in place, routine mining and processing of resources, and initial investigation of unknown or particularly hazardous environments.

But, of course, it's always humans who really do exploring. Their enhanced capabilities to interpret, synthesize, and adapt are not properties of machines. The human brain consists of a compact supercomputer that is both programmable and *re*programmable by training, experience, and preceding observations. Human eyes form a high-resolution stereo optical system of great dynamic range, and their integration with the brain provides capabilities for discovery and interpretation far beyond those of robotic cameras. Human hands constitute a still underutilized, highly dexterous bio-mechanical system that, when integrated with the brain and eyes, is unmatched in scientific potential.

Most important, humans react spontaneously to the exploration environment, bringing instant creativity to bear on any new circumstance, opportunity, or problem. My discovery of the critically important orange pyroclastic glass during the *Apollo 17* mission illustrates this fact in spades. Ongoing studies of that one small sample of unique material continue to inform our quest to understand the Moon's origin and early history, knowledge that may well shed light on the origins and early history of Earth.

And if any of the human field geologists reading these pages had been exploring Mars in place of the current, wonderfully successful pair of rovers, they would have quickly resolved any doubt about the nature of the intriguing layered outcrops in Eagle Crater. As Steven Squyres also has noted, they would have explored far more terrain, rapidly integrating their findings into the complex and mysterious picture of Martian history. Human explorers would have been able to quickly test the hypotheses that *Opportunity* was exploring sedimentary rocks laid down by wind or by standing water, confirming or rejecting either possibility in favor of a better one based on close observation and judicious use of a rock hammer.

As important as cost-effective scientific exploration of new worlds is for understanding nature and our place in it, other facets of future human activity in space override even these arguments. Among them is the natural urge, common to all species, to expand accessible habitats and thus enhance a species' prospects for long-term survival. In this regard, settlement of the Moon and Mars clearly is now technically feasible for our species. Moreover, the refinement of the technology that will be required for settlement will bring into being the additional technology that will have economic and environmental benefits here on Earth, including lunar energy resources and material recycling and conservation.

Finally, I believe that if the American people continue to encourage their elected leaders to truly look to the future, a special benefit from deep space exploration and settlement exists if Americans lead that activity. When humans return to the Moon and go to Mars, the transplantation of our institutions of freedom to those first human settlements on other worlds eventually will create new free societies that will sprout and grow. This is our special gift and our special obligation to the future.

—HARRISON H. SCHMITT, PhD
Lunar Module Pilot, *Apollo 17*
Former U.S. Senator (NM)
Chairman, Interlune-InterMars Initiative, Inc.
Adjunct Professor of Engineering, University of Wisconsin–Madison

# ACKNOWLEDGMENTS

THIS IS THE BOOK I originally set out to write in the fall of 1997, after the end of the Mars Pathfinder mission. It was to be the people story. With a draft of chapters, I naively set out to find a literary agent to sell it to a publisher, but since I wasn't planning on giving up my day job, I asked the agents I contacted about possible coauthors. I got nowhere. Selling a new Mars mission to NASA was easier.

My breakthrough came following a talk a group of us gave at an event hosted by Pritchett and Associates. Price Pritchett is a dynamic and very personable man who writes and publishes a wide variety of books specifically tuned to the business issue of the day. Following our talk, I approached Price with a question about getting a book published. At the time, he offered what wisdom he could, but only a few weeks later he called with the suggestion that the two of us write a business book together, which he would publish. It was a wonderfully natural collaboration, and we accomplished it in a Pathfinder-like three months from concept to finished product with the insightful help of aide-de-camp extraordinaire Stephanie Snyder. But while it was a wonderful product, very well designed and executed, it was still not the book I'd always wanted to write.

Again random chance stepped in when Price and I realized we had a common friend in the publishing business, Adrian Zackheim, then of HarperBusiness. Adrian wanted a business book written by Price and me. We talked seriously about it, but Price, after assessing his other responsibilities, said he couldn't commit to such a project. Adrian did offer an excellent alternate coauthor, Bill Simon, whom he knew could work well with a "hands-on" guy like me. Bill and I did work well together and produced a quality business-oriented book, *High-Velocity Leadership*. But the book I wanted to write was still out there. So out of inspiration (or desperation!) I called Rick Berman, producer of the *Star Trek* television series and movies. I'd met Rick on a *Star Trek* set after he'd invited members of the Pathfinder team to visit and share our common interest in "going where no one had gone before." I thought a guy like Rick must know some good writers, and he immediately recommended I contact a husband-and-wife team with the distinctive handle of Judith and Garfield Reeves-Stevens. Judy's and Gar's enthusiasm for a joint project was matched by their excellent relationship and shared excitement of executive editor Margaret Clark at Pocket Books. Showing Gar, Judy, and Margaret around JPL was like taking the proverbial kids to the proverbial candy store, and off we went to finally write the book I had hoped for from the beginning. Then came the failures of the Mars '98 missions. The dreams of a faster, cheaper, and better space program and a new book deal immediately came to a high-g stop, if not crash.

But you know what they say about an idea whose time has come. The geometry of the planets and laws of celestial mechanics allowed a twenty-six-month hiatus for NASA and JPL to assess the problems and mistakes of the Mars '98 missions. With the recognition that 2003 represented a very good opportunity to send hardware back to Mars, the enterprise resumed with new direction and under new rules. And so did this book. While Judy and Gar and I remained friends and talked about future work, I really had my doubts about whether or not the book would be revived along with NASA's new plans for Mars, but due to the good offices of Margaret Clark we were back at it. Gar and Judy are as enthusiastic a pair of space cadets as I've ever had the pleasure to work with. One of the qualities that I speak about in my non-JPL speaking and consulting business is Daniel Goleman's concept of "emotional intelligence." EQ can be summed up as the qualities of energy, drive, commitment, enthusiasm, and optimism that, in fact, play a bigger role in the success of most endeavors than all other measures of technical ability combined. Although I know of no scale that measures EQ, I know Judy and Gar are at the top of it.

Of course I must acknowledge my friends and colleagues at NASA and Caltech's Jet Propulsion Laboratory. As should be clear in these pages, JPL is a place of enormous power and capability brought about by a culture that has grown up from the very beginnings of the space program and has managed to keep its edge as the business has evolved from *Explorer I* to simple fly-by machines to complex landers invented on demand, all under an increasingly harsher and less forgiving spotlight. The space business is engineering at its most creative, and nobody does it better than the men and women from all over the world who have assembled to conquer deep space at NASA's JPL.

I want to extend my thanks to the many people whom we interviewed for this book. Without exception, whether recounted here or not, they contributed to the richness of *Going to Mars*. I also want to thank Dave Spencer for checking that I got the mission design sidebars right, Ed Jorgenson for verifying cost numbers, and Mary-Sue O'Brien (of JPL's Document Review Office) for providing the required review and being the first person to critique the manuscript.

How do you appropriately thank a person who has led the scientific understanding of another world? Humbly. I could not have been happier when Harrison Schmidt agreed to provide the afterword for this book. His perspective provides a perfect nexus of the past and the future of human exploration. I also greatly appreciate JPL Director Dr. Charles Elachi for his support of this book, for his inspiring foreword, and for me personally in my career at JPL. He personifies the best of JPL in his technical brilliance, powerful leadership, and most especially his attention to the people of JPL.

Finally, I must say how much I love my daughters, Alicia and Jenna. For while they seem to be totally jaded about the work their mother (my ex-wife, Catherine Cagle, a talented and successful engineer and manager at JPL) and I do, I believe, in the recesses of their hearts, that they actually do understand and appreciate how cool their parents' jobs really are.

—Brian Muirhead

SCIENCE AND SCIENCE FICTION are constant sources of inspiration for each other, so it's entirely fitting that Rick Berman provided the spark that brought this book to life by introducing us to Brian Muirhead. As executive producer, writer, and creator of the *Star Trek* television series and movies, Rick inspires the scientists and engineers of NASA and JPL to make real the spacefaring future with which he entertains the world. We thank him for inspiring Brian and us to work together, and we look forward to the day the spacedocks of Utopia Planitia on Mars launch their first real starship.

Seven years is a long time to work on a book, especially when that seven-year span includes the loss of two Mars missions in late 1999, which to many in the publishing world also meant the loss of an audience. This book survived that dark time for NASA and JPL because of the unwavering support of our editor, Margaret Clark, and the head of Pocket Books' media division, Scott Shannon. Special appreciation also goes to Elizabeth Braswell, whose early support of this project helped make it a reality, and to our agent, Martin Shapiro of Shapiro-Lichtman, who shepherded the project through all its trials.

Margaret Clark's passion for this book is equal to ours and Brian's, and her involvement in every aspect, particularly its design and organization, reflects her deeply appreciated professionalism and enthusiasm for the subject, and her authors. This is every bit as much her book as it is ours.

Also at Pocket, once again we are the grateful recipients of designer Richard

Oriolo's fine eye and ability to bring order to a disparate collection of words and pictures. We thank him for transforming our stacks of unruly manuscript into a collection of pages that truly are a joy to behold.

But none of the efforts of our publishing team would mean anything without the kindness and support of the people of JPL. Going to Mars is a tough business, and schedules are such that regular hours and weekends are rare luxuries. Yet everyone in these pages found personal time to speak with us, in the euphoria of Pathfinder, the depression of Mars '98, the tension of MER, and finally, the renewed euphoria of *Spirit*'s and *Opportunity*'s perfect landings. We are especially appreciative of JPL's director, Dr. Charles Elachi, who not only supported this project but provided the foreword. And—just as Peter Theisinger reminds us that when we see seven people on the press-conference stage we must remember that they represent the four thousand people who made MER possible—for every person quoted in these pages, five more spoke with us, informing the writing of this book. Even at twice the length, we could not include everyone at JPL who contributed to our work, but we thank them all.

Our thanks also to Harrison Schmidt for providing the afterword. It is incredibly inspiring to have an explorer and scientist from the first phase of humanity's personal exploration of other worlds point the way to the next phase.

We also thank our friends at the Planetary Society for their help in tracking down rare photos, and for running Planetfest '97, where we started our own journey to Mars by watching the first images come back from *Pathfinder* in the company of hundreds of members of the world's largest and most successful space advocacy organization. (Please check the resources section of this book to see how you can join!)

Thanks also to Carol Monroe of Hasbro for helping us secure the ad for the toy that inspired our coauthor. What wonders will we see in years to come from the children playing with the Hasbro toys of today?

On a personal note, we also thank Doug Netter and Steve Ecclesine, who lived through the last year of this book with us, and whose enthusiasm for the project remains a constant source of inspiration. The journey won't be complete until you're on Mars with us.

And finally, our thanks to Brian Muirhead, who welcomed two "space cadets" into the fold to share the excitement (and the frustration) of what it means to actually go where no one has gone before.

While working with him on this book, we benefited enormously from his insights into what motivates people and how the raw talents of hundreds of highly individualistic and creative engineers can be harnessed to achieve a single goal. It's no surprise to us that Brian's first two books were about leadership, because he excels at it, and it has been a privilege for us to see that expertise in action—because it's exactly that kind of expertise that *will* get us to Mars.

We're glad to have made such a good friend along the way.

—J&G Reeves-Stevens

# RESOURCES

As a government agency responsible to American taxpayers, NASA provides a range of resources and services to keep the public informed about its many accomplishments and its plans for the future, as well as offering a wealth of educational materials to schools. Here are some of the opportunities available for extending the content of this book.

## TOURS

### JET PROPULSION LABORATORY

JPL is located in Pasadena, California, which, depending on traffic, is less than half an hour from major Los Angeles attractions like Universal Studios. To tour the Lab, advance reservations are necessary, and the reservations must be made by speaking directly with a Public Services Office representative—no fax, voice messages, or e-mail allowed. As of this writing, tour reservations are booked solid for six months, so plan early!

Tours generally run from two to two and a half hours, and usually include a multimedia presentation about the Lab entitled "Welcome to Outer Space," which provides an overview of JPL's activities and accomplishments. Guests may also visit the von Kármán Visitor Center, the Space Flight Operations Facility, and the Spacecraft Assembly Facility.

For reservations, contact:

Public Services Office
Mail Stop 186-113
Jet Propulsion Laboratory
4800 Oak Grove Drive
Pasadena, CA 91109
Phone: 818-354-9314
Fax: 818-393-4641

### KENNEDY SPACE CENTER

The Kennedy Space Center at Cape Canaveral, Florida, offers an entertaining and informative mix of tours, activities, and attractions, about a forty-five-minute drive from Walt Disney World. Witnessing an actual launch is a thrilling event, and is something we recommend to everyone.

The best way to find out about all KSC has to offer, as well as check the dates for scheduled launches, is to log on to www.kennedyspacecenter.com.


## ON THE INTERNET

NASA maintains a comprehensive Web site that serves as an entry point to almost every other NASA center, including JPL. Access it at www.nasa.gov.

JPL can be reached directly at www.jpl.nasa.gov for all the latest updates on its many missions, including those to Mars and Saturn.

For more specific information on the Mars Pathfinder mission, two versions of the original Web site still exist. Version One presents the Web site as it was on the day *Pathfinder* landed—July 4, 1997. Version Two presents the Web site at the mission's completion.

The original site can be found at http://mars.jpl.nasa.gov/MPF/index0.html.

The final site can be found at http://mars.jpl.nasa.gov/MPF/index1.html, and includes links to many other Mars missions and resources.

For the Mars Exploration Rover missions, still under way as this book goes to press, log onto http://marsrovers.jpl.nasa.gov/home/index.html. *Every* image taken by *Spirit* and *Opportunity* is available at this site, along with a complete collection of press releases detailing the rovers' impressive accomplishments and groundbreaking discoveries.

If you've ever wanted to plan your own rover operations on Mars, be sure to go to http://mars.telascience.org/home/ and download your own copy of MAESTRO—a public version of the software package used to operate *Spirit* and *Opportunity*, complete with actual images and science observations obtained from the real Mars rovers.

Finally, to take in the big picture, be sure to visit Malin Space Science Systems at http://www.msss.com. At that site, more than 150,000 images of Mars taken from orbit are available for browsing, and true Mars buffs can even have a hand in suggesting which parts of Mars should be imaged next.

## THE PLANETARY SOCIETY

The authors respectfully suggest that if you've read this far in the book, you would love to be a member of the Planetary Society! The Society was founded in 1980 by Carl Sagan, Bruce Murray, and Louise Friedman to encourage the exploration of our solar system and the search for extraterrestial life. It's a nonprofit, nongovernmental organization, funded by dues and donations from individuals around the world. With more than 100,000 members from over 140 countries, it's the largest space advocacy group on Earth, and membership is open to everyone who's interested in the Society's mission.

The Planetary Society encourages all spacefaring nations to explore other worlds; provides public information and supports educational activities about the exploration of the solar system and the search for extraterrestrial life; and supports and funds innovative and novel research and development projects that can seed future projects of planetary exploration. Members also receive a bimonthly magazine, *The Planetary Report*, with the latest news about space exploration.

Find out more on the Internet, at http://www.planetary.org.

## THE AUTHORS

And finally, for more information about Brian Muirhead, log on to www.velocity-works.com. For Judith and Garfield Reeves-Stevens, visit www.reeves-stevens.com.